Lecture Notes in Computer Science   8065

Commenced Publication in 1973
Founding and Former Series Editors:
Gerhard Goos, Juris Hartmanis, and Jan van Leeuwen

Ngoc-Thanh Nguyen (Ed.)

Transactions on Computational Collective Intelligence XI

 Springer

Volume Editor

Ngoc-Thanh Nguyen
Wrocław University of Technology
Institute of Informatics
Wybrzeże Wyspiańskiego 27, 50-370 Wrocław, Poland
E-mail: ngoc-thanh.nguyen@pwr.edu.pl

ISSN 0302-9743 (LNCS) e-ISSN 1611-3349 (LNCS)
ISSN 2190-9288 (TCCI)
ISBN 978-3-642-41775-7 e-ISBN 978-3-642-41776-4
DOI 10.1007/978-3-642-41776-4

Springer Heidelberg New York Dordrecht London

Library of Congress Control Number: 2013951518

CR Subject Classification (1998): I.2.11, I.2, H.3, I.6, I.5, F.1

Typesetting: Camera-ready by author, data conversion by Scientific Publishing Services, Chennai, India

Printed on acid-free paper

Springer is part of Springer Science+Business Media (www.springer.com)

Transactions on Computational Collective Intelligence XI

Preface

Welcome to the ninth volume of *Transactions on Computational Collective Intelligence* (TCCI). This is the third issue in 2013, the third year of TCCI activities.

This volume of TCCI includes nine interesting and original papers selected via a peer-review process. The first paper, entitled "Taming Complex Beliefs" by Barbara Dunin-Kęplicz and Andrzej Szałas, presents a novel formalization of beliefs in multi-agent systems. This is to bridge the gap between idealized logical approaches to modeling beliefs and their actual implementations. In this approach, a novel semantics reflecting these stages is provided. In the second paper, "Ideal Chaotic Pattern Recognition Is Achievable: The Ideal-M-AdNN – Its Design and Properties," the authors Ke Qin and B. John Oommen address the relatively new field of designing a chaotic pattern recognition (PR) system. The authors have shown that the modified Adachi neural network, when tuned appropriately, is capable of demonstrating ideal PR capabilities, and that it is able to switch to being periodic whenever it encounters patterns with which it was trained. The next paper, "A Framework for an Adaptive Grid Scheduling: An Organizational Perspective" by Inès Thabet, Chihab Hanachi, and Khaled Ghédira, provides a grid conceptual model that identifies the concepts and entities involved in the cooperative scheduling activity. The authors use this model to define a typology of adaptation including perturbing events and actions to undertake in order to adapt. A simulator and an experimental evaluation have also been realized to demonstrate the feasibility of the approach. In the fourth paper entitled "Data Extraction from Online Social Networks Using Application Programming Interface in a Multi Agent System Approach" the authors, Ruqayya Abdulrahman, Daniel Neagu, D.R.W. Holton, Mick Ridley, and Yang Lan, present the extension of their previous work that implemented an online social network retrieval system (OSNRS) to decentralize the retrieving information process from online social networks. The novelty of OSNRS is its ability to monitor the user profiles continuously. In the fifth paper titled "Cooperatively Searching Objects Based on Mobile Agents" by Takashi Nagata, Munehiro Takimoto, and Yasushi Kambayashi, a framework for controlling multiple robots connected by communication networks is presented. The authors propose a multiple robot control approach based on mobile agents for searching targets as one of the effective examples. They have shown that it can be extended to other more practical examples, or be used as an element of a real application because of its simplicity. The sixth paper entitled "Agent-Based Optimisation of VoIP Communication", by Drago Žagar and Hrvoje Očevčić, presents a model for simple management and optimization of VoIP quality of service. The model is within the framework of VoIP communication optimization based on simple measurement information using agent architecture for VoIP QoS management. In the next paper titled "Towards Rule Interoperability: Design of Drools Rule Bases Using the XTT2 Method," Krzysztof Kaczor, Krzysztof Kluza, and Grzegorz J. Nalepa present a unified and formalized method for knowledge interchange for the most common rule languages. The approach involves three levels of interoperability abstraction: a semantic level, a model level, and an environment level. The eighth paper, "Artificial Immune System for Forecasting Time Series with Multiple Seasonal Cycles" by Grzegorz Dudek, presents a new immune-inspired univariate method for forecasting time series with multiple seasonal periods. This method is based on the patterns of time series seasonal sequences. In the last paper, "Machine Ranking of 2-Uncertain Rules Acquired from Real Data," Beata Jankowska and Magdalena Szymkowiak propose a method for ranking 2-uncertain rules obtained from real data. The method can facilitate creating high-quality diagnostic rule-based systems with uncertainty. The authors have acquired the rules from aggregate attributive data with set values, by means of their semantic integration.

TCCI is a peer-reviewed and authoritative journal dealing with the working potential of computational collective intelligence (CCI) methodologies and applications, as well as emerging issues of interest to academics and practitioners. The research area of CCI has been growing significantly in recent years and we are very thankful to everyone within the CCI research community who has supported the *Transactions on Computational Collective Intelligence* and its affiliated events including the International Conferences on Computational Collective Intelligence (ICCCI). Its last event (ICCCI 2012) was held in Ho Chi Minh City, Vietnam, in November 2012. The next event will be held in Craiova, Romania, in September 2013. It is a tradition that after each ICCCI event, we invite authors of selected papers to extend their manuscripts and submit them for publication in TCCI.

We would like to thank all the authors, Editorial Board members, and the reviewers for their contributions to TCCI. We express our sincere gratitude to all of them. Finally, we would also like to express our thanks to the LNCS editorial staff of Springer with Alfred Hofmann, who support the TCCI journal.

May 2013   Ngoc-Thanh Nguyen

Transactions on Computational Collective Intelligence

This Springer journal focuses on research in computer-based methods of computational collective intelligence (CCI) and their applications in a wide range of fields such as the Semantic Web, social networks, and multi-agent systems. It aims to provide a forum for the presentation of scientific research and technological achievements accomplished by the international community.

The topics addressed by this journal include all solutions of real-life problems for which it is necessary to use computational collective intelligence technologies to achieve effective results. The emphasis of the papers is on novel and original research and technological advancements. Special features on specific topics are welcome.

Table of Contents

Taming Complex Beliefs*

Barbara Dunin-Kęplicz[1,2] and Andrzej Szałas[1,3]

[1] Institute of Informatics, University of Warsaw
Banacha 2, 02-097 Warsaw, Poland
{keplicz,andrzej.szalas}@mimuw.edu.pl
[2] Institute of Computer Science
Polish Academy of Sciences, Warsaw, Poland
[3] Department of Computer and Information Science, Linköping University
SE-581 83 Linköping, Sweden

Abstract. A novel formalization of beliefs in multiagent systems has recently been proposed by Dunin-Kęplicz and Szałas. The aim has been to bridge the gap between idealized logical approaches to modeling beliefs and their actual implementations. Therefore the stages of belief acquisition, intermediate reasoning and final belief formation have been isolated and analyzed. In conclusion, a novel semantics reflecting those stages has been provided. This semantics is based on the new concept of epistemic profile, reflecting agent's reasoning capabilities in a dynamic and unpredictable environment. The presented approach appears suitable for building complex belief structures in the context of incomplete and/or inconsistent information. One of original ideas is that of epistemic profiles serving as a tool for transforming preliminary beliefs into final ones. As epistemic profile can be devised both on an individual and a group level in analogical manner, a uniform treatment of single agent and group beliefs has been achieved.

In the current paper these concepts are further elaborated. Importantly, we indicate an implementation framework ensuring tractability of reasoning about beliefs, propose the underlying methodology and illustrate it on an example.

1 Beliefs in Multiagent Systems

During the past years awareness has been intensively investigated both from the theoretical as well as from the practical perspective. Its importance manifested itself especially in the context of cooperating teams of agents or other mixed groups in the context of intelligent, autonomous systems. In multiagent systems, agents' awareness is typically expressed in terms of different (combinations of) beliefs about

- the environment;
- an agent itself;
- other agents/groups involved.

Such beliefs are built using various forms of observations, communication and reasoning [2,12,13,15,36]. Existing modern, fine-grained logic-based approaches typically

---

* Supported by the Polish National Science Centre grants 2011/01/B/ST6/02769 and 6505/B/T02/2011/40.

exploit rather subtle (combinations of) multi-modal logics [16,18,19,20]. Unfortunately this usually leads to high complexity of reasoning that is unacceptable from the point of view of their implementation and use. In fact, the underlying semantical structures are rather abstract and hardly reflect the way beliefs are acquired and finally formed. To make it even worse, in many applications one needs to take into account relevant features of perception, including:

- limited accuracy of sensors and other devices;
- restrictions on time and other resources affecting measurements;
- unfortunate combinations and unpredictability of environmental conditions;
- noise, limited reliability and failure of physical devices.

In multiagent systems during belief formation initial and intermediate beliefs are confronted with other beliefs originating from a variety of sources. The resulting beliefs can then substantially deviate from the initial ones. Moreover, there might still exist areas of agents' ignorance and inconsistencies. A low quality of information does not waive agents' responsibility of decision making. Therefore, reducing the areas of ignorance and inconsistencies is vital. In modern systems this can be accomplished in many different ways, including

- a variety of reasoning methods;
- belief exchange by communication;
- belief fusion;
- supplementary observations.

Apparently there is no guarantee to acquire the whole necessary information and/or to resolve all inconsistencies. Information may still remain partly unknown and/or inconsistent. Such situations may be sorted out by the use of

- paraconsistent models allowing for inconsistencies and lack of information;
- nonmonotonic reasoning techniques for completing missing information and resolving inconsistencies.

However, both paraconsistent and nonmonotonic reasoning, in their full generality, are intractable [5,17,21,25]. This naturally restricts their use in multiagent systems and calls for a shift in perspective. In [10] we proposed a novel framework for flexibly modeling beliefs of heterogenous agents, inspired by knowledge representation and deductive database techniques.

The key abstraction is that of *epistemic profiles* reflecting agent's individual reasoning capabilities. In short, epistemic profile defines a schema in which an agent reasons, deals with conflicting information and deals with its ignorance. These skills are achievable by combining various forms of reasoning, including belief fusion, disambiguation of conflicting beliefs or completion of lacking information. This rich repertoire of available methods enables for heterogeneity of agents' reasoning characteristics. More importantly, the same approach may be applied to groups of agents or even more complex mixed groups, allowing for uniform treatment of these, essentially different, cases.

In the current paper we show how the framework of [10] can serve as a basis for actual implementations and ensure tractability of reasoning. This is achieved by representing sets of preliminary and final beliefs as well as epistemic profiles using the deductive databases machinery of the 4QL query language [22,23,24,35].

The rest of this paper is structured as follows. In Section 2 our approach to structuring beliefs is outlined and motivated. Next, in Section 3, formal syntax and semantics of the basic language are defined. The pragmatics of its use in multiagent systems is discussed in Section 4. Section 5 is devoted to distributed beliefs. Implementation issues and complexity are addressed in Section 6. Finally, Section 7 concludes the paper.

The current paper is an extended and revised version of papers [10,11].

2 Structuring Beliefs

In the sequel, belief formation by agents will be unveiled. In the idealized logical approaches to agency, this paradigmatic part of agents' activity seems to be, at least partly, neglected. We analyze and model this process from the very beginning, that is from agents' perception and other kinds of basic beliefs.

The basis for the framework is formed by semantical structures reflecting the processes of an agent's belief acquisition and formation. Namely, an agent starts with *constituents*, i.e., sets of beliefs acquired by:

- perception;
- expert supplied knowledge;
- communication with other agents;
- other ways.

Next, the constituents are transformed into *consequents* according to the agent's *individual epistemic profile*.

While building a multiagent system, lifting beliefs to the group or even more complex level is substantial. As regards belief formation, it would be perfect to reach a conceptual compatibility between individual and group cases. Assuming that *the group epistemic profile* is set up, analogical individual and group procedures are then applicable for defining belief fusion methods, where:

- consequents of group members become constituents at the group level;
- such constituents are further transformed into group consequents.

Observe that this way various perspectives of agents involved are taken into consideration and merged. Moreover, we use the same underlying semantical structures for groups and individuals. The only requirement is that all epistemic profiles of complex structures are fixed. This way a uniform approach applies to groups of groups of agents or to mixed groups of individuals and other complex topologies.

Example 2.1. Consider an agent equipped with a sensor platform for detecting air pollution and two different sensors for measuring the noise level. The agent has also some information about the environment, including places in the neighborhood, etc. The task is to decide whether conditions in the tested position are healthy.

It is natural to consider, among others, three constituents:

- C_p gathering beliefs about air pollution at given places, in terms of $P(x, y)$ indicating the pollution level y at place x, where $y \in \{low, moderate, high\}$;
- C_n gathering beliefs about noise level at given places, in terms of $N_i(x, y)$ indicating the noise level y at place x, as measured by a sensor $i \in \{1, 2\}$, where $y \in \{low, moderate, high\}$;
- C_e gathering information about the environment in terms of $Cl(x, y)$ indicating that place x is close to a place characterized by y, where $y \in \{pollutive, noisy, neutral\}$.

For example, we may have:

$$C_p = \{P(a, low)\}, \quad C_n = \{N_1(a, high)\},$$
$$C_e = \{\neg Cl(a, noisy), Cl(a, neutral), Cl(a, pollutive)\}.$$

Note that we have no information from the second noise sensor (no literal $N_2()$ is given) and somehow inconsistent information as to the pollution level (C_p indicates low level, but according to C_e the agent is close to a pollutive location). Also there is an implicit disagreement between $N_1(a, high)$ appearing in C_e and $\neg Cl(a, noisy)$ appearing in C_e, which may be caused by a defective information source.

Based on constituents, the agent has to decide whether the situation is healthy or not (and include the thus obtained belief to the set of consequents). For example, the agent may accept

$$F = \{\neg S(a, healthy), S(a, healthy)\}$$

as its consequent, i.e., it may have inconsistent beliefs about the issue whether the situation at place a is healthy.   ◁

3   Syntax and Semantics

Inconsistency in common-sense reasoning attracted recently many logicians. To model inconsistencies, a commonly used logic is the four-valued logic proposed in [4]. However, as discussed, e.g., in [8,37], this approach is problematic in many applications. In particular, disjunction and conjunction deliver results which can be misleading for more classically oriented users.

On the other hand, our approach is strongly influenced by ideas underlying the 4QL query language [22,23,24] which does not share such problems. 4QL is a rule-based DATALOG⌐⌐-like query language that provides simple, yet powerful constructs for expressing nonmonotonic rules reflecting, among others, default reasoning, autoepistemic reasoning, defeasible reasoning, local closed world assumption, etc. [22]. 4QL enjoys tractable query computation and captures all tractable queries. Therefore, 4QL is a natural implementation tool creating a space for a diversity of applications. To our knowledge a paraconsistent approach to beliefs has mainly been pursued in the context of belief revision [26,30]. However, these papers use formalisms substantially different from ours (like models based on criteria and rationality indexes [30] or relevant logic [26]).

Most of the approaches to modeling beliefs in logic start with variants of Kripke structures [12,15,16,18,19,34,36], where possible worlds are total and consistent. This

Table 1. Truth tables for ∧, ∨, → and ¬ (see [37,22])

∧	f	u	i	t	∨	f	u	i	t	→	f	u	i	t	¬	
f	f	f	f	f	f	f	u	i	t	f	t	t	t	t	f	t
u	f	u	u	u	u	u	u	i	t	u	t	t	t	t	u	u
i	f	u	i	i	i	i	i	i	t	i	f	f	t	f	i	i
t	f	u	i	t	t	t	t	t	t	t	f	f	t	t	t	f

creates natural problems in modeling certain types of ignorance. For example, it is currently unknown whether, say, the Riemann's hypothesis is true. Therefore, to model this situation, we would have to create (at least) two possible worlds: one where the hypothesis is true and one where it is false. However, one of them would become inconsistent. In order to address such problems, modal frames involving non-standard worlds are considered [31,38]. However, our solution is simpler and leads to a substantial reduction of complexity.

In what follows all <u>sets are finite</u> except for sets of formulas.

We deal with the classical first-order language over a given vocabulary without function symbols. We assume that *Const* is a fixed set of constants, *Var* is a fixed set of variables and *Rel* is a fixed set of relation symbols.

Definition 3.1. *A literal is an expression of the form $R(\bar{\tau})$ or $\neg R(\bar{\tau})$, with τ being a sequence of arguments, $\bar{\tau} \in (Const \cup Var)^k$, where k is the arity of R. Ground literals over Const, denoted by $\mathcal{G}(Const)$, are literals without variables, with all constants in Const. If $\ell = \neg R(\bar{\tau})$ then $\neg\ell \stackrel{\text{def}}{=} R(\bar{\tau})$.* ◁

Though we use the classical first-order syntax, the presented semantics substantially differs from the classical one. Namely,

- truth values $\mathsf{t}, \mathsf{i}, \mathsf{u}, \mathsf{f}$ (true, inconsistent, unknown, false) are explicitly present;[1]
- the semantics is based on sets of ground literals rather than on relational structures.

This allows one to deal with the lack of information as well as inconsistencies. As 4QL is based on the same principles, it can immediately be used as the implementation tool.

The semantics of propositional connectives is summarized in Table 1. Observe that definitions of ∧ and ∨ reflect minimum and maximum w.r.t. the ordering:

$$\mathsf{f} < \mathsf{u} < \mathsf{i} < \mathsf{t}, \tag{1}$$

as advocated, e.g., in [7,22,37]. Such a truth ordering seems to be quite natural. It indicates how "true" a given proposition is. The value f indicates that the proposition is definitely not true, u admits a possibility that the proposition is true, i shows that there

---

[1] For simplicity we use the same symbols to denote truth constants and corresponding truth values.

is at least one witness/evidence indicating the truth of the proposition, and finally, t expresses that the proposition is definitely true.

Note that (1) linearizes the truth ordering of [4], where u and i are incomparable. This linearization is compatible with knowledge ordering of [4] where u < i.

According to [35], the pragmatics of disjunction should include the following principles:

- disjunction is true only when at least one of its operands is true;
- disjunction is false only when all its operands are false;

and the pragmatics of conjunction:

- conjunction is true only when all its operands are true;
- conjunction is false only when at least one of its operands is false.

The implication $\rightarrow$ is a four-valued extension of the classical implication. It is motivated and discussed in [22,23,37,35]. Observe that implication can only be t or f. Implication

$$premisses \rightarrow conclusion$$

reflects the following principles [35]:

- truth or falsity of the conclusion can only be deduced when premises are true;
- when premises are inconsistent, conclusion should be inconsistent, too;
- false or unknown premises do not participate in deriving new conclusions.

Remark 3.2. It is worth emphasizing that [35]:

- when one restricts truth values to $\{t, f\}$ then connectives defined in Table 1 become equivalent to their counterparts in classical propositional logic;
- when one restricts truth values to $\{t, u, f\}$ or to $\{t, i, f\}$ then conjunction, disjunction and negation become respectively their counterparts in Kleene three-valued logic K_3 with the third (non-classical) value meaning *undetermined* and in Priest logic P_3 [30], where the third value receives the meaning *paradoxical*.[2]   ◁

Let $v : Var \longrightarrow Const$ be a *valuation of variables*. For a literal ℓ, by $\ell(v)$ we understand the ground literal obtained from ℓ by substituting each variable x occurring in ℓ by constant $v(x)$.

Definition 3.3. The *truth value* of a literal ℓ w.r.t. a set of ground literals L and valuation v, denoted by $\ell(L, v)$, is defined as follows:

$$\ell(L, v) \stackrel{\text{def}}{=} \begin{cases} t & \text{if } \ell(v) \in L \text{ and } (\neg\ell(v)) \notin L; \\ i & \text{if } \ell(v) \in L \text{ and } (\neg\ell(v)) \in L; \\ u & \text{if } \ell(v) \notin L \text{ and } (\neg\ell(v)) \notin L; \\ f & \text{if } \ell(v) \notin L \text{ and } (\neg\ell(v)) \in L. \end{cases}$$

◁

---

[2] The only difference between K_3 and P_3 is that only *true* is designated in K_3, while in P_3 both *true* and *paradoxical* are.

Table 2. Semantics of first-order formulas

- if α is a literal then $\alpha(L, v)$ is defined in Definition 3.3;
- $(\neg\alpha)(L, v) \stackrel{\text{def}}{=} \neg(\alpha(L, v))$, where $\neg$ at the righthand side of equality is defined in Table 1;
- $(\alpha \circ \beta)(L, v) \stackrel{\text{def}}{=} \alpha(L, v) \circ \beta(L, v)$, where $\circ \in \{\vee, \wedge, \rightarrow\}$;
- $(\forall x\alpha(x))(L, v) \stackrel{\text{def}}{=} \min_{a \in Const} \{(\alpha_a^x)(L, v)\}$, where min is the minimum w.r.t. ordering (1);
- $(\exists x\alpha(x))(L, v) \stackrel{\text{def}}{=} \max_{a \in Const} \{(\alpha_a^x)(L, v)\}$, where max is the maximum w.r.t. ordering (1).

Example 3.4. Consider the situation described in Example 2.1 and let $v(x) = low$. Then, for example, $P(a, x)(C_p, v) = \mathbf{t}$ and $P(a, x)(C_e, v) = \mathbf{u}$. $\lhd$

For a formula $\alpha(x)$ with a free variable x and $c \in Const$, by $\alpha(x)_c^x$ we understand the formula obtained from α by substituting all free occurrences of x by c. Definition 3.3 is extended to all formulas in Table 2, where α and β denote first-order formulas, v is a valuation of variables, L is a set of ground literals, and the semantics of propositional connectives appearing at righthand sides of equivalences is given in Table 1.

Let us now define belief structures based on sets of literals. In this context the concept of an epistemic profile is the key abstraction involved in belief formation.

If S is a set then by $\text{FIN}(S)$ we understand the set of all finite subsets of S.

Let us now define the concepts of belief structures and epistemic profiles which are central to our approach.

Definition 3.5. Let $\mathbb{C} \stackrel{\text{def}}{=} \text{FIN}(\mathcal{G}(Const))$ be the set of all finite sets of ground literals over the set of constants *Const*. Then:

- by a *constituent* we understand any set $C \in \mathbb{C}$;
- by an *epistemic profile* we understand any function $\mathcal{E} : \text{FIN}(\mathbb{C}) \longrightarrow \mathbb{C}$;
- by a *belief structure over an epistemic profile* $\mathcal{E}$ we mean $\mathcal{B}^{\mathcal{E}} = \langle \mathcal{C}, F \rangle$, where:
  - $\mathcal{C} \subseteq \mathbb{C}$ is a nonempty set of constituents;
  - $F \stackrel{\text{def}}{=} \mathcal{E}(\mathcal{C})$ is the *consequent* of $\mathcal{B}^{\mathcal{E}}$. $\lhd$

Example 3.6. For the Example 2.1,
$\mathcal{C} = \{C_p, C_n, C_e\}$ and $F = \{\neg S(a, healthy), S(a, healthy)\}$,
so $\mathcal{E}$ is any function of the signature required in Definition 3.5 such that $\mathcal{E}(\mathcal{C}) = F$. $\lhd$

Note that constituents and consequents contain ground literals only. Of course, they can be defined using advanced theories or deductive database technologies. Therefore, if one wants to express beliefs as expressions more complex than just literals, it can be done. In Definition 3.5 we do not restrict representation of constituents or consequents. It may be of arbitrary complexity. The only requirement is that representations used should finally return finite sets of ground literals. The same applies to epistemic profiles. However, when tractability is to be achieved, such representations should be restricted to algorithms running in deterministic polynomial time. Our choice is to find representations based on deductive database technologies that ensure tractability and capture all polynomially computable queries. This way we guarantee both tractability and possibility of expressing all epistemic profiles and belief structures constructible in deterministic polynomial time.

Definition 3.7. Let $\mathcal{E}$ be an epistemic profile. The *truth value of formula* α w.r.t. belief structure $\mathcal{B}^{\mathcal{E}} = \langle \mathcal{C}, F \rangle$ and valuation v, denoted by $\alpha(\mathcal{B}^{\mathcal{E}}, v)$, is defined by:[3]

$$\alpha(\mathcal{B}^{\mathcal{E}}, v) \stackrel{\text{def}}{=} \alpha(\bigcup_{C \in \mathcal{C}} C, v).$$

◁

Example 3.8. Consider again the situation described in Example 2.1. Let $v(x) = low$ and the belief structure $\mathcal{B}^{\mathcal{E}}$ be as described in Example 3.6. Then,

$$\bigcup_{C \in \mathcal{C}} C = \{P(a, low), N_1(a, high), \neg Cl(a, noisy), Cl(a, neutral), Cl(a, pollutive)\}.$$

Therefore, e.g., $(P(a,x) \wedge N_1(a,x))(\mathcal{B}^{\mathcal{E}}, v) = \mathsf{u}$ and $(P(a,x) \vee N_1(a,x))(\mathcal{B}^{\mathcal{E}}, v) = \mathsf{t}$. Observe that truth of $N_1(a, high)$ does not automatically imply falsity of $N_1(a, low)$. ◁

To express beliefs, we extend the language with operator $\mathrm{Bel}()$ standing for beliefs. The truth table for $\mathrm{Bel}()$ is:

$$\mathrm{Bel}(\mathsf{t}) \stackrel{\text{def}}{=} \mathsf{t}, \quad \mathrm{Bel}(\mathsf{i}) \stackrel{\text{def}}{=} \mathsf{i}, \quad \mathrm{Bel}(\mathsf{u}) \stackrel{\text{def}}{=} \mathsf{f}, \quad \mathrm{Bel}(\mathsf{f}) \stackrel{\text{def}}{=} \mathsf{f}. \tag{2}$$

We say that a formula is $\mathrm{Bel}()$-*free* if it contains no occurrences of the $\mathrm{Bel}()$ operator.

Definition 3.9. Let $\mathcal{E}$ be an epistemic profile. The *truth value of formula* α *w.r.t. belief structure* $\mathcal{B}^{\mathcal{E}} = \langle \mathcal{C}, F \rangle$ *and valuation* v, denoted by $\alpha(\mathcal{B}^{\mathcal{E}}, v)$, is defined as follows:

– clauses for propositional connectives and quantifiers are as in Table 2;
– when α is $\mathrm{Bel}()$-free then:
  • $\alpha(\mathcal{B}^{\mathcal{E}}, v)$ is defined by Definition 3.7;
  • $\mathrm{Bel}(\alpha)(\mathcal{B}^{\mathcal{E}}, v) \stackrel{\text{def}}{=} \mathrm{Bel}(\alpha(F, v))$, where the truth value $\alpha(F, v)$ is defined in Table 2 and $\mathrm{Bel}()$ applied to a truth value is defined by (2);
– when $\mathrm{Bel}()$ operators are nested in α then $\alpha(\mathcal{B}^{\mathcal{E}}, v)$ is evaluated starting from the innermost occurrence of $\mathrm{Bel}()$, which is then replaced by the obtained truth value, etc. ◁

Example 3.10. For the belief structure $\mathcal{B}^{\mathcal{E}}$ introduced in Example 2.1 (see also Example 3.6) and $v(x) = low$, we have:

$$\big(P(a,x) \wedge \mathrm{Bel}(P(a,x) \vee \mathrm{Bel}(S(a, healthy)))\big)(\mathcal{B}^{\mathcal{E}}, v) =$$
$$\mathsf{t} \wedge \mathrm{Bel}(P(a, low) \vee \mathrm{Bel}(\mathsf{i})) = \mathsf{t} \wedge \mathrm{Bel}(P(a, low) \vee \mathsf{i}) = \mathsf{t} \wedge \mathrm{Bel}(\mathsf{u} \vee \mathsf{i}) =$$
$$\mathsf{t} \wedge \mathrm{Bel}(\mathsf{i}) = \mathsf{t} \wedge \mathsf{i} = \mathsf{i}.$$

◁

---

[3] Since $\bigcup_{C \in \mathcal{C}} C$ is a set of ground literals, $\alpha(\mathcal{S}, v)$ is well-defined by Table 2.

One can easily verify the following proposition.

Proposition 3.11. *For any formula* α, *belief structure* $\mathcal{B}^{\mathcal{E}}$ *and valuation of variables* v:

$$\left(\neg\mathrm{Bel}(\mathsf{f})\right)\left(\mathcal{B}^{\mathcal{E}},v\right)=\mathsf{t}; \tag{3}$$

$$\left(\mathrm{Bel}(\alpha)\rightarrow\mathrm{Bel}(\mathrm{Bel}(\alpha))\right)\left(\mathcal{B}^{\mathcal{E}},v\right)=\mathsf{t}; \tag{4}$$

$$\left(\neg\mathrm{Bel}(\alpha)\rightarrow\mathrm{Bel}(\neg\mathrm{Bel}(\alpha))\right)\left(\mathcal{B}^{\mathcal{E}},v\right)=\mathsf{t}. \tag{5}$$

$\lhd$

Observe that the above formulas express the classical properties of beliefs: (3) is the axiom **D**, (4) and (5) are axioms **4** and **5**, expressing positive and negative introspection. Note that modal logic **KD45**, based on these axioms, is typically used to model beliefs in multiagent systems. Furthermore, there are belief structures, where the following axiom **T**, distinguishing knowledge and beliefs, does not have to be true:

$$\mathrm{Bel}(\alpha)\rightarrow\alpha. \tag{6}$$

This follows from the fact that given a belief structure $\langle\mathcal{C},F\rangle$, $\mathrm{Bel}(\alpha)$ evaluates α in F while α itself is evaluated in $\bigcup\mathcal{C}$.

Let us also note that the following axiom:

$$\neg\left(\mathrm{Bel}(\alpha)\wedge\mathrm{Bel}(\neg\alpha)\right), \tag{7}$$

sometimes replacing axiom (3) is not always true under our semantics. For example, when α is i then formula (7) is i. Axioms (3) and (7) are equivalent in the context of **KD45**. However, this is no longer the case in our semantics as we allow agents to have inconsistent beliefs.

Observe, however, that in our semantics $\mathrm{Bel}(\alpha\vee\beta)$ has always the same truth value as $\mathrm{Bel}(\alpha)\vee\mathrm{Bel}(\beta)$. This is caused by the fact that we assume that any epistemic profile delivers a single consequent (a single "world"). This is closer to the intuitionistic understanding of disjunction than to the classical one. In applications requiring that $\mathrm{Bel}(\alpha\vee\beta)$ does not force $\mathrm{Bel}(\alpha)\vee\mathrm{Bel}(\beta)$ one has to allow more than one consequent in Definition 3.5 introducing belief structures.

4 Pragmatics

4.1 Individual Beliefs

Agents can acquire knowledge about other agents' beliefs via communication and observation. In contrast to many existing approaches, we do not assume that an agent entering a group changes its beliefs. However, group beliefs prevail over individual ones. For example, when two agents cooperate, they may have certain beliefs as a group, but do not have to share them as individuals. Such a perspective usually results in a substantial improvement of complexity.

When the group is dismissed, agents continue to act according to their individual beliefs. These can be revised to reflect information acquired during cooperation. In our approach such revisions are delayed until the group disintegrates. Actually, in everyday life we frequently face similar situations. A group member does not need to share beliefs of group leaders, but still has to obey their commands. As group beliefs are typically built upon individual ones, immediate revisions of group members' beliefs could force revision at the group level. In reasonable cases one can expect that this process would converge to a fixpoint, but this is not guaranteed. Therefore, in certain situations infinite loops in belief revisions could occur.

When agents cooperate, a specific group is, possibly implicitly, created, including an epistemic profile fitting the entire situation. This is where belief fusion methods adequate for the group in question occur. In general, any interaction between agents leads to the creation of a (possibly virtual) group with a specific epistemic profile. We can naturally model this process in the proposed framework.

Let $\{Ag_i \mid i = 1, \ldots, n\}$ be a *set of agents*. To model individual beliefs we introduce belief operators $\mathrm{Bel}_i(\alpha)$, for $i = 1, \ldots, n$. As usually, the formula $\mathrm{Bel}_i(\alpha)$ expresses that agent Ag_i believes in α. To define the semantics of $\mathrm{Bel}_i(\alpha)$ operators we assume that for $i = 1, \ldots, n$, $\mathcal{E}_i$ is an epistemic profile of agent Ag_i and $\mathcal{B}^{\mathcal{E}_i} = \langle \mathcal{C}_i, F_i \rangle$ is a belief structure of agent Ag_i.

Definition 4.1. Let $\bar{\mathcal{B}} = \{\mathcal{B}^{\mathcal{E}_i} \mid i = 1, \ldots, n\}$ be a tuple of belief structures. The *truth value of formula* α *w.r.t.* $\bar{\mathcal{B}}$ *and valuation* v *w.r.t. agent* Ag_i, denoted by $\alpha(i, \bar{\mathcal{B}}, v)$, is defined as follows:

- clauses for propositional connectives and quantifiers are as in Table 2;
- when α is $\mathrm{Bel}()$-free then:
  - $\alpha(i, \bar{\mathcal{B}}, v)$ is defined as $\alpha(\mathcal{B}^{\mathcal{E}_i}, v)$ in the sense of Definition 3.7;
  - $\mathrm{Bel}_j(\alpha)(i, \bar{\mathcal{B}}, v) \stackrel{\text{def}}{=} \mathrm{Bel}(\alpha(F_j, v))$, where the truth value $\alpha(F_i, v)$ is defined in Table 2 and $\mathrm{Bel}()$ applied to a truth value is defined by (2);
- when $\mathrm{Bel}_j()$ operators are nested in α then $\alpha(i, \bar{\mathcal{B}}, v)$ is evaluated starting from the innermost occurrence of $\mathrm{Bel}()$, which is then replaced by the obtained truth value, etc.   ◁

Example 4.2. When agent Ag_k evaluates formula $\big(r \vee \mathrm{Bel}_i(\mathrm{Bel}_j(p) \wedge q)\big)$ w.r.t. v then:

- r is evaluated w.r.t. F_k and v;
- p in $\mathrm{Bel}_j(p)$ is evaluated w.r.t. F_j and v;
- q in $\mathrm{Bel}_i(\mathrm{Bel}_j(p) \wedge q)$ is evaluated w.r.t. F_i and v.   ◁

4.2 Group Beliefs

A group of agents, say $G = \{Ag_{i_1}, \ldots, Ag_{i_k}\}$, has its *group belief structure* $\mathcal{B}^{\mathcal{E}_G} = \langle \mathcal{C}_G, F_G \rangle$, where $\mathcal{C}_G = \{F_{i_1}, \ldots, F_{i_k}\}$. Thus, consequents of group members become constituents of a group. The group then builds group beliefs via its epistemic profile $\mathcal{E}_G$, e.g. by adjudicating beliefs of group members, and reaches its consequent F_G (see Figure 1).

To express properties of group beliefs we extend the language by allowing operators $\mathrm{Bel}_G(\alpha)$, where G is a group of agents.

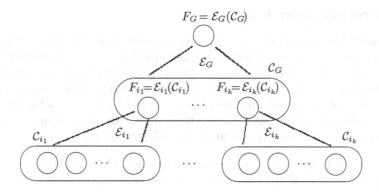

Fig. 1. The architecture of individual and group beliefs

Let $\{Ag_i \mid i = 1, \ldots, n\}$ be a *set of agents* and $\{G_j \mid j = n + 1, \ldots m\}$ be a set of groups of agents. To define the semantics of $\mathrm{Bel}_G(\alpha)$ operators we extend Definition 4.1 by assuming that for $l = n + 1, \ldots, m$, $\mathcal{E}_l$ is an epistemic profile of group G_l and $\mathcal{B}^{\mathcal{E}_l} = \langle \mathcal{C}_l, F_l \rangle$ is a belief structure of group G_l. We therefore have a tuple of belief structures $\bar{\mathcal{B}} = \{\mathcal{B}^{\mathcal{E}_l} \mid l = 1, \ldots, m\}$, where for $i = 1, \ldots, n$, $\mathcal{B}^{\mathcal{E}_i}$ is a belief structure of agent Ag_i and for $j = n + 1, \ldots, m$, $\mathcal{B}^{\mathcal{E}_j}$ is a belief structure of group G_j.

Since groups are dealt with exactly as agents, given a tuple of belief structures $\bar{\mathcal{B}} = \{\mathcal{B}^{\mathcal{E}_l} \mid l = 1, \ldots, m\}$, the truth value of formula α w.r.t. $\bar{\mathcal{B}}$ and valuation v w.r.t. agent Ag_i (respectively, group G_i), is defined exactly as in Definition 4.1, assuming that indices $1, \ldots, n$ refer to agents and indices $n + 1, \ldots, m$ refer to groups.

4.3   Other Complex Beliefs

The same way, beliefs of groups involving other groups may be formed. For example, a surveillance group of robots G_s may join a rescue team of robots G_r making a larger group $G_{s,r}$. Then the consequents of G_s and G_r become constituents of $G_{s,r}$. Furthermore, such groups can become parts of other, more complex groups, and so on. The underlying methods for forming group beliefs on the top of group members' beliefs are typically highly application- and context-dependent.

To express beliefs of such groups we extend the language with $\mathrm{Bel}_{\mathbb{G}}()$ operators, where $\mathbb{G}$ may contain individual agents, groups of agents, groups of groups of agents, etc. Since such complex groups, when formed, are equipped with belief structures like in the case of groups consisting of agents only, the semantics of $\mathrm{Bel}_{\mathbb{G}}()$ operators is given by an immediate adaptation of the semantics of groups.

Note that, due to complexity reasons, it is reasonable to assume that only formed groups are equipped with belief structures and epistemic profiles. When a group does not exist, we assume that its belief structure $\mathcal{B}$ is "empty", i.e., $\mathcal{B} = \langle \mathcal{C}, F \rangle$ with $\mathcal{C} = \{\emptyset\}$ (that is, $\mathcal{C}$ consists of a single set being the empty set) and $F = \emptyset$. Note that all queries supplied to this structure return the value u.

4.4   Querying Belief Structures

Traditional deductive databases are mainly based on the classical logic [1]. Belief operators are rather rarely considered in such contexts (but see, e.g., [9,28,29]). In our approach belief operators relate formulas to consequents which are sets of ground literals. Therefore, rather than with possible worlds, we always deal with sets of ground literals present in considered structures. Namely, in order to find out what are actual beliefs of agents, groups of agents, etc., the mechanism based on querying belief structures is applicable. For example one can ask the following queries:

$\mathrm{Bel}(\exists X(S(X, healthy)))$ – is it believed that there is a healthy place?
$\mathrm{Bel}(\forall X(S(X, healthy)))$ – is it believed that all places are healthy?

Belief fusion requires gathering beliefs of different agents. For example, the following query:

$$\mathrm{Bel}_1(\exists X(S(X, healthy))) \wedge \mathrm{Bel}_2(\exists X(S(X, healthy))), \tag{8}$$

allows us to check whether agents $Ag1$ and $Ag2$ believe that there is a place where the situation is healthy. Formula (8) contains no free variables, so the query returns a truth value. Of course, beliefs of these agents do not have to refer to the same place. If one intends to verify whether there is a place believed to be healthy by both agents simultaneously, then the query should rather be formulated as:

$$\exists X(\mathrm{Bel}_1(S(X, healthy)) \wedge \mathrm{Bel}_2(S(X, healthy))). \tag{9}$$

Using query:

$$\mathrm{Bel}_1(\forall X(S(X, healthy))) \vee \mathrm{Bel}_2(\forall X(S(X, healthy))) \tag{10}$$

one can ask whether at least one of agents believes that all places are healthy.
   On the other hand, query:

$$\forall X(\mathrm{Bel}_1(S(X, healthy)) \vee \mathrm{Bel}_2(S(X, healthy))) \tag{11}$$

expresses the fact that every place is believed to be healthy by at least one agent.
   When formulas used as queries contain free variables, queries return tuples together with an information whether a given tuple makes the query t, f or i (tuples making the query u are not returned) – see [35]. For example, the query:

$$\mathrm{Bel}_1(S(X, healthy)) \wedge \mathrm{Bel}_2(S(X, healthy)) \tag{12}$$

asks for values of X such that both agents believe that X is healthy. One can get an answer that a place a is healthy, a place b is not healthy, that it is inconsistent that a place c is healthy. Since there is no information about other places, it is unknown whether these places are healthy or not.
   Such, possibly rather complex, queries are naturally used in designing epistemic profiles. Let us also note, that in multiagent settings they provide a powerful mechanism for deciding which actions to perform.

5   Distributed Beliefs

In contemporary intelligent distributed systems, like multiagent systems, we typically deal with many heterogenous information sources. They independently deliver information (e.g. percepts), expressed in terms of beliefs, on various aspects of a recent situation. Depending on the context and the goal of the reasoning process, different beliefs need to be fused in order to achieve more holistic judgement of the situation. Apparently, this information fusion may be realized in various ways. Let us now take a closer look at this formal process.

Distributed information sources naturally introduce four truth values [4,10,22,23,24]. On the other hand, in real-world distributed problem solving, lack of knowledge and inconsistent beliefs are to be resolved at some point. More precisely, in the case of lacking knowledge, we need to complete missing information at the objective level, while in the case of inconsistencies we do this at a meta-level, for example, by verifying which information sources deliver false information. This knowledge can then be used for a better setup or calibration of sensors and other data sources, as well as diagnostic systems detecting malfunctioning devices. When this is impossible, especially in time-critical systems, commonsense reasoning methods can be of help as they generally characterize typical situations [6,21,25]. Among these methods (local) closed world assumption, default reasoning, autoepistemic reasoning and defeasible reasoning are of primary importance. Again, 4QL supports such forms of reasoning.

As we shall discuss in the next sections, using our approach one can achieve a tractable model of distributed belief fusion, as well as an implementation framework of distributed belief fusion via epistemic profiles.

6   Towards Implementation

As indicated before, we apply reasoning over databases rather than over general theories. Such an approach reflects the reality of intelligent systems and significantly reduces the complexity of reasoning, typically from at least exponential to deterministic polynomial time, no matter what type of reasoning is used. This substantial complexity gain is achieved by using 4QL, a query language [22,23,24] which enjoys tractable query computation and captures all tractable queries. Belief structures, when implemented in 4QL, can be considered as sets of ground literals generated by facts and rules. Therefore, one can tractably query belief structures using such query languages as first-order queries, fixpoint queries or 4QL queries.

6.1   The Main Idea

The main idea is illustrated in Figure 2. Namely, we propose to implement epistemic profiles via an intermediate layer consisting of *derivatives*, where each derivative is a finite set of ground literals. Intuitively, derivatives represent intermediate belief fusion results or, in other words, intermediate views on the situation in question. Importantly, such a structure allows us to implement belief fusion in a highly distributed manner.

Example 6.1. One can consider two derivatives:

- D_p – for deciding the pollution level;
- D_n – for deciding the noise level.

Such derivatives should result from reasoning patterns defined by the corresponding epistemic profile of the considered agent. For example, these derivatives may be:

$$D_p = \{P(a, moderate)\}, \quad D_n = \{N(a, high)\}.$$

Based on the contents of D_p and D_n, the agent has to decide whether the situation is healthy or not (and include it in its set of consequents F).   ◁

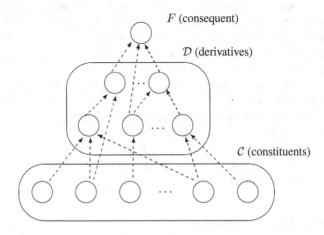

Fig. 2. Implementation framework for belief structures and epistemic profiles. Arrows indicate belief fusion processes.

6.2 Implementation Tool: 4QL

There are several languages designed for programming BDI agents (for a survey see, e.g., [27]). However, none of these approaches directly addresses belief formation, in particular nonmonotonic/defeasible reasoning techniques. Our choice is therefore 4QL, a DATALOG$^{\neg\neg}$-like query language. It supports a modular and layered architecture, and provides a tractable framework for many forms of rule-based reasoning both monotonic and nonmonotonic. As the underpinning principle, openness of the world is assumed, which may lead to the lack of knowledge. Negation in rule heads, expressing negative conclusions, may lead to inconsistencies. As indicated in [22], to reduce the unknown/inconsistent zones, *modules* and *external literals* provide means for:

- the application-specific disambiguation of inconsistent information;
- the use of (Local) Closed World Assumption;
- the implementation of various forms of nonmonotonic and defeasible reasoning.

To express nonmonotonic/defeasible rules we apply modules as well as external literals, originally introduced in [22]. Importantly, different modules can be distributed among different agents participating in the reasoning process.

In the sequel *Mod* denotes the set of *module names*.

Definition 6.2. An *external literal* is an expression of one of the forms:

$$M.R, -M.R, M.R \text{ IN } T, -M.R \text{ IN } T, \tag{13}$$

where $M \in Mod$ is a module name, R is a positive literal, '$-$' stands for negation and $T \subseteq \{\mathsf{f}, \mathsf{u}, \mathsf{i}, \mathsf{t}\}$. For literals (13), module M is called the *reference module*.   ◁

The intended meaning of "$M.R$ IN T" is that the truth value of $M.R$ is in the set T. External literals allow one to access values of literals in other modules. If R is not defined in the module M then the value of $M.R$ is assumed to be u.

Assume a strict tree-like order $\prec$ on *Mod* dividing modules into layers. An external literal with reference module M_1 may appear in rule bodies of a module M_2, provided that $M_1 \prec M_2$.[4]

Definition 6.3. By a *rule* we mean any expression of the form:

$$\ell :- b_{11}, \ldots, b_{1i_1} \mid \ldots \mid b_{m1}, \ldots, b_{mi_m}. \tag{14}$$

where ℓ is a literal, $b_{11}, \ldots, b_{1i_1}, \ldots, b_{m1}, \ldots, b_{mi_m}$ are literals or external literals, and ',' and '$\mid$' abbreviate conjunction and disjunction, respectively.

Literal ℓ is called the *head* of the rule and the expression at the righthand side of :− in (14) is called the *body* of the rule.   ◁

Rules of the form (14) are understood as implications:

$$((b_{11} \wedge \ldots \wedge b_{1i_1}) \vee \ldots \vee (b_{m1} \wedge \ldots \wedge b_{mi_m})) \rightarrow \ell,$$

where it is assumed that the empty body takes the value t in any set of literals.

By convention, facts are rules with the empty body. For example, a fact '$P(a)$.' is an abbreviation for the rule '$P(a)$:−.' .

Definition 6.4. Let a set of constants, *Const*, be given. A set of ground literals L with constants in *Const* is a *model of a set of rules S* iff each ground instance of each rule of S (understood as implication) obtains the value t in L.   ◁

The semantics of 4QL is defined via well-supported models generalizing the idea presented in [14]. Intuitively, a model is *well-supported* if all derived literals are supported by a reasoning grounded in facts. It appears that for any set of rules there is a unique well-supported model and it can be computed in polynomial time. For details see [24].

Remark 6.5. One can further extend 4QL without losing its tractability by allowing arbitrary first-order formulas in bodies of rules. This allows one to directly implement queries like those considered in Section 4.4. For details of such an extension see [35].◁

---

[4] Observe that layers generalize the concept of stratification of DATALOG¬ queries [23] (for definition of stratification see, e.g., [1]).

6.3 Implementing Belief Structures in 4QL

Implementation of tractable belief structures and epistemic profiles is now relatively easy. Namely,

- epistemic profiles can be implemented by the use of derivatives;
- every constituent, derivative and consequent can be implemented as a separate 4QL module.

The hierarchical structure of derivatives makes it possible to use the layered architecture of modules, as required in 4QL. External literals allow to access information to create beliefs on the basis of perhaps still preliminary beliefs already obtained.

Let $\{\mathcal{E}_i \mid i = 1, \ldots, n\}$ be epistemic profiles of agents $Ag_1, \ldots, Ag_n$ and let $\mathcal{B}_i^{\mathcal{E}} = \langle \mathcal{C}_i, F_i \rangle$ be the agents' belief structures over these epistemic profiles. Assume an agent Ag_k ($1 \leq k \leq n$) is asked a query. We have the following two cases:

1. when a formula expressing (a part of) a query is not within the scope of a $\mathrm{Bel}()$ operator then we evaluate it in the database obtained as the union of constituents $\bigcup_{C \in \mathcal{C}_k} C$ (according to Definition 3.7);

2. when a formula has the form $\mathrm{Bel}_j(\alpha)$ ($1 \leq j \leq n$), it is evaluated in F_j (considered as a database).

Example 6.6. Consider queries (8)–(11) (Section 4.4). If $Ag_1.S, Ag_2.S$ respectively refer to relation S included in the set of consequents of Ag_1's and Ag_2's belief structures then queries (8)–(11) can be expressed by:

$$\exists X(Ag_1.S(X, healthy)) \land \exists X(Ag_2.S(X, healthy)),$$
$$\exists X(Ag_1.S(X, healthy) \land Ag_2.S(X, healthy)),$$
$$\forall X(Ag_1.S(X, healthy)) \lor \forall X(Ag_2.S(X, healthy)),$$
$$\forall X(Ag_1.S(X, healthy) \lor Ag_2.S(X, healthy)),$$

where $Ag_i.S$ indicates that the value of S is taken from consequents of agent Ag_i. ◁

Open source interpreters [32,33] of 4QL are available via 4ql.org. Also, a commercial implementation of 4QL is being recently developed by NASK.[5]

6.4 Exemplary Implementation

Consider now the scenario outlined in Examples 2.1 and 6.1. Exemplary modules corresponding to constituents, derivatives and consequents are shown in Tables 3–5, respectively.[6]

It is important to note that well-supported models are sets of literals. Thus relational or deductive databases technology can be used to query them (see also Section 4.4).

In fact, four logical values, external literals and modular architecture distinguish 4QL from many other approaches (for a survey see, e.g., [3]). 4QL modules are structured

---

[5] http://www.nask.pl/nask_en/
[6] We use the Inter4QL self-explanatory syntax – for details see [33].

Table 3. Modules corresponding to constituents considered in Example 2.1

```
module Cp:              module Cn:              module Ce:
  domains:                domains:                domains:
    literal level.          literal level.          literal level.
    literal place.          literal place.          literal place.
  relations:              relations:              relations:
    P(place, level).        N1(place, level).       Cl(place, type).
  facts:                    N2(place, level).     facts:
    P(a, low).            facts:                    -Cl(a, noisy).
end.                        N1(a, high).            Cl(a,neutral).
                                                    Cl(a,pollutive).
                                                  end.
```

into layers. According to syntax of 4QL, the lowest layer represents monotonic reasoning, while higher ones allow the user to provide (nonmonotonic) rules for disambiguating inconsistencies and completing lacking information. This is achieved by explicitly referring to logical values via external literals.

6.5 Complexity

Theoretically, any belief structure (also augmented with derivatives) can be of exponential size w.r.t. the number of literals involved. In applications they may be dynamically generated as agents appear or groups are formed. However, at a given timepoint we can assume that their size is always reasonable, as it reflects available resources. Also the number of groups may, in theory, be of exponential size w.r.t. the number of agents. Again, in a given application we have a limited number of groups and only these are equipped with belief structures, as discussed in [10] and in this paper itself.

Let $|Const| = k$, let n be the number of agents and let m be the number of groups. Further, let the size of all belief structures involved be bounded by $f(k, n, m)$. Then we have the following theorem which follows from the tractability of 4QL [22,23,24].

Theorem 6.7. *If belief structures and queries are implemented using the 4QL query language then the time complexity of computing queries is deterministic polynomial in* $f(k, n, m)$. ◁

Let us emphasize again that in practice one can safely assume that $f(k, n, m)$ is bounded by available resources, including time, memory and physical devices.

Theorem 6.7 holds also when 4QL is replaced by any query language with polynomially bounded complexity of computing queries. However, 4QL captures all polynomially computable queries [23], so when 4QL is used, we also have another theorem.

Theorem 6.8. *Any polynomially constructed belief structure can be implemented using 4QL.* ◁

It is worth emphasizing that Theorem 6.8 shows that 4QL is a sufficient language that serves our purposes. One can argue that the same applies to any query language which captures deterministic polynomial time. However, in contrast to other languages,

Table 4. Modules corresponding to derivatives considered in Example 6.1

```
module Dp:
  domains:
    literal level.
    literal place.
  relations:
    P(place, level).
  rules:
    P(X, moderate):-  Cp.P(X, low) IN {TRUE, UNKNOWN},
                      Ce.Cl(X, pollutive) IN {TRUE, UNKNOWN}.
    ...
end.

module Dn:
  domains:
    literal level.
    literal place.
  relations:
    N(place, level).
  rules:
    N(X,Y):- Cn.N1(X,Y), Cn.N2(X,Y) IN {TRUE, UNKNOWN} |
             Cn.N1(X,Y) IN {TRUE, UNKNOWN}, Cn.N2(X,Y).
    ...
end.
```

Table 5. Module corresponding to consequents considered in Example 2.1

```
module F:
  domains:
    literal characteristics.
    literal place.
  relations:
    S(place, characteristics).
  rules:
    -S(X, healthy):- Dn.N1(X, high), Dn.N2(X, high).
    S(X, healthy):- Cp.P(X, moderate),
                    Cn.N1(X, low) IN {TRUE, UNKNOWN}.
    ...
end.
```

4QL provides simple, but powerful tools for direct expression of a wide spectrum of reasoning techniques, including nonmonotonic ones, and allowing one to handle inconsistencies.

Observe also that agents' and groups' belief structures and epistemic profiles typically match some patterns reflecting agents' types or groups' organizational structures and cooperation procedures. Therefore, in practice, one can expect belief structures to be generated on the basis of a library of patterns, much like in object-oriented programming dynamic objects are generated on the basis of static classes, developed during the system's design phase. Of course, the number of such patterns does not change during system execution, so can be bound by a constant.

7 Conclusions

In this paper we differentiated agents' characteristics via individual and group epistemic profiles, reflecting agents' reasoning capabilities. This abstraction tool permits both to flexibly define the way an agent (a group) reasons and to reflect the granularity of reasoning. The presented pragmatic framework to beliefs suits real-world applications that often are not easy to formalize. In particular, it allows for natural handling of inconsistencies and gaps in beliefs by using paraconsistent and nonmonotonic reasoning.

Moreover, our approach permits a uniform modeling of individual and group beliefs, where group is a generic concept consisting of individual agents, groups of agents, groups of groups of agents, etc. Importantly, the assumed layered architecture underlying the framework allows one to avoid costly revisions of agents' beliefs when they join a group. This is especially important when paradigmatic agent interactions are considered. Cooperation, coordination and communication is naturally modeled by creating a group and forming group beliefs to achieve a common informational stance. What sort of structure it is and how this influences agents' individual beliefs is a matter of design decisions. Our approach ensures both the heterogeneity of agents involved and flexibility of group level reasoning patterns.

Most significantly, we have indicated 4QL as a tool to implement all epistemic profiles and belief structures constructible in deterministic polynomial time. We have also shown a natural methodology to obtain such implementations. One can then query implemented belief structures in a tractable manner, which provides a rich but still pragmatic reasoning machinery. To the best of our knowledge, such tractability of reasoning about beliefs has not been achieved yet. Also, nonmonotonic/defeasible reasoning techniques are easily expressible in 4QL, ensuring both richness and flexibility of implemented epistemic profiles.

References

1. Abiteboul, S., Hull, R., Vianu, V.: Foundations of Databases. Addison-Wesley Pub. Co. (1996)
2. Ågotnes, T., Alechina, N.: Full and relative awareness: A decidable logic for reasoning about knowledge of unawareness. In: Proc. of TARK, pp. 6–14. ACM Press (2007)
3. Alferes, J.J., Pereira, L.M.: Reasoning with Logic Programming. LNCS, vol. 1111. Springer, Heidelberg (1996)
4. Belnap, N.D.: A useful four-valued logic. In: Epstein, G., Dunn, J.M. (eds.) Modern Uses of Many Valued Logic, pp. 8–37. Reidel (1977)
5. Béziau, J.-Y., Carnielli, W., Gabbay, D.M. (eds.): Handbook of Paraconsistency. College Publications (2007)
6. Brewka, G.: Non-Monotonic Reasoning: Logical Foundations of Commonsense. Cambridge University Press (1991)
7. de Amo, S., Pais, M.S.: A paraconsistent logic approach for querying inconsistent databases. International Journal of Approximate Reasoning 46, 366–386 (2007)
8. Dubois, D.: On ignorance and contradiction considered as truth-values. Logic Journal of the IGPL 16(2), 195–216 (2008)
9. Dunin-Kęplicz, B., Nguyen, L.A., Szałas, A.: Tractable approximate knowledge fusion using the Horn fragment of serial propositional dynamic logic. Int. J. Approx. Reasoning 51(3), 346–362 (2010)

10. Dunin-Kęplicz, B., Szałas, A.: Epistemic Profiles and Belief Structures. In: Jezic, G., Kusek, M., Nguyen, N.-T., Howlett, R.J., Jain, L.C. (eds.) KES-AMSTA 2012. LNCS, vol. 7327, pp. 360–369. Springer, Heidelberg (2012)
11. Dunin-Kęplicz, B., Szałas, A.: Paraconsistent Distributed Belief Fusion. In: Proc. IDC 2012, 6th International Symposium on Intelligent Distributed Computing. SCI, vol. 446, pp. 59–69. Springer, Heidelberg (2012)
12. Dunin-Kęplicz, B., Verbrugge, R.: Teamwork in Multi-Agent Systems. A Formal Approach. John Wiley & Sons, Ltd. (2010)
13. Dunin-Kęplicz, B., Verbrugge, R.: Awareness as a vital ingredient of teamwork. In: Stone, P., Weiss, G. (eds.) Proc. of AAMAS 2006, pp. 1017–1024 (2006)
14. Fages, F.: Consistency of Clark's completion and existence of stable models. Methods of Logic in Computer Science 1, 51–60 (1994)
15. Fagin, R., Halpern, J.: Belief, awareness, and limited reasoning. Artificial Intelligence 34(1), 39–76 (1988)
16. Fagin, R., Halpern, J.Y., Moses, Y., Vardi, M.Y.: Reasoning About Knowledge. The MIT Press (2003)
17. Gottlob, G.: Complexity results for nonmonotonic logics. Journal of Logic and Computation 2(3), 397–425 (1992)
18. Hintikka, J.: Knowledge and Belief. Cornell University Press (1962)
19. Kraus, S., Lehmann, D.: Knowledge, belief and time. Theoretical Computer Science 58, 155–174 (1988)
20. Laux, A., Wansing, H. (eds.): Knowledge and Belief in Philosophy and Artificial Intelligence. Akademie Verlag, Berlin (1995)
21. Łukaszewicz, W.: Non-Monotonic Reasoning - Formalization of Commonsense Reasoning. Ellis Horwood (1990)
22. Małuszyński, J., Szałas, A.: Living with inconsistency and taming nonmonotonicity. In: de Moor, O., Gottlob, G., Furche, T., Sellers, A. (eds.) Datalog 2010. LNCS, vol. 6702, pp. 384–398. Springer, Heidelberg (2011)
23. Małuszyński, J., Szałas, A.: Logical foundations and complexity of 4QL, a query language with unrestricted negation. Journal of Applied Non-Classical Logics 21(2), 211–232 (2011)
24. Małuszyński, J., Szałas, A.: Partiality and Inconsistency in Agents Belief Bases. In: Agents and Multi-agent Systems – Technologies and Applications. IOS Press (to appear, 2013)
25. Marek, V.W., Truszczyński, M.: Nonmonotonic Logic. Springer (1993)
26. Mares, E.D.: A paraconsistent theory of belief revision. Erkenntnis 56, 229–246 (2002)
27. Mascardi, V., Demergasso, D., Ancona, D.: Languages for programming BDI-style agents: an overview. In: Corradini, F., De Paoli, F., Merelli, E., Omicini, A. (eds.) WOA 2005 - Workshop From Objects to Agents, pp. 9–15 (2005)
28. Nguyen, L.A.: Foundations of modal deductive databases. Fundamenta Informaticae 79(1-2), 85–135 (2007)
29. Nguyen, L.A.: Constructing finite least Kripke models for positive logic programs in serial regular grammar logics. Logic Journal of the IGPL 16(2), 175–193 (2008)
30. Priest, G.: Paraconsistent belief revision. Theoria 67(3), 214–228 (2001)
31. Rantala, V.: Impossible worlds semantics and logical omniscience. Acta Philosophica Fennica 35, 106–115 (1982)
32. Sobczyk, Ł.: 4QL Runner developer description (2012),
    http://4ql.org/wp-content/uploads/2012/10/4qlRunner.pdf
33. Spanily, P.: The Inter4QL interpreter (2012),
    http://4ql.org/wp-content/uploads/2012/10/inter4ql.pdf

34. Stalnaker, R.: On logics of knowledge and belief. Philosophical Studies 128, 169–199 (2006)
35. Szałas, A.: How an agent might think. Logic Journal of the IGPL 21(3), 515–535 (2013)
36. van Ditmarsch, H.P., van der Hoek, W., Kooi, B.P.: Dynamic Epistemic Logic. Springer (2007)
37. Vitória, A., Małuszyński, J., Szałas, A.: Modeling and reasoning with paraconsistent rough sets. Fundamenta Informaticae 97(4), 405–438 (2009)
38. Wansing, H.: A general possible worlds framework for reasoning about knowledge and belief. Studia Logica 49, 523–539 (1990)

Ideal Chaotic Pattern Recognition Is Achievable: The Ideal-M-AdNN - Its Design and Properties

Ke Qin[1,*] and B. John Oommen[2,**]

[1] School of Computer Science and Engineering
University of Electronic Science & Technology of China
Chengdu, China, 610054
qinke@uestc.edu.cn
[2] School of Computer Science, Carleton University,
1125 Colonel By Dr., Ottawa, ON, K1S 5B6, Canada
oommen@scs.carleton.ca

Abstract. This paper deals with the relatively new field of designing a *Chaotic* Pattern Recognition (PR) system. The benchmark of such a system is the following: First of all, one must be able to train the system with a set of "training" patterns. Subsequently, as long as there is no testing pattern, the system must be chaotic. However, if the system is, thereafter, presented with an unknown testing pattern, the behavior must ideally be as follows. If the testing pattern is not one of the trained patterns, the system must continue to be chaotic. As opposed to this, if the testing pattern is truly one of the trained patterns (or a noisy version of a trained pattern), the system must switch to being *periodic*, with the specific trained pattern appearing periodically at the output. This is truly an ambitious goal, with the requirement of switching from chaos to periodicity being the most demanding. Some related work has been done in this regard. The Adachi Neural Network (AdNN) [1–5] has properties which are pseudo-chaotic, but it also possesses *limited* PR characteristics. As opposed to this, the Modified Adachi Neural Network (M-AdNN) proposed by Calitoiu *et al* [6], is a fascinating NN which has been shown to possess the required periodicity property desirable for PR applications. However, in this paper, we shall demonstrate that the PR properties claimed in [6] are not as powerful as originally reported. Indeed, the claim of the authors of [6] is true, in that it resonates periodically for *trained* input patterns. But unfortunately, the M-AdNN also resonates for unknown patterns and produces these unknown patterns at the output periodically. However, we describe how the parameters of the M-AdNN for its weights, steepness and external inputs, can be specified so as to yield a new NN, which we shall refer to as the Ideal-M-AdNN.

* The work of this author was partially supported by grant No. 2012HH0003 and 9140A17060411DZ02.
** *Chancellor's Professor*; *Fellow : IEEE* and *Fellow : IAPR*. This author is also an *Adjunct Professor* with the University of Agder in Grimstad, Norway. The work of this author was partially supported by NSERC, the Natural Sciences and Engineering Research Council of Canada.

N.T. Nguyen (Ed.): Transactions on CCI XI, LNCS 8065, pp. 22–51, 2013.
© Springer-Verlag Berlin Heidelberg 2013

Using a rigorous Lyapunov analysis, we shall analyze the chaotic properties of the Ideal-M-AdNN, and demonstrate its chaotic characteristics. Thereafter, we shall verify that the system is also truly chaotic for untrained patterns. But most importantly, we demonstrate that it is able to *switch to being periodic* whenever it encounters patterns with which it was trained. Apart from being quite fascinating, as far as we know, the theoretical and experimental results presented here are both unreported and novel. Indeed, we are not aware of any NN that possesses these properties!

Keywords: Chaotic Neural Networks, Chaotic Pattern Recognition, Adachi-like Neural Networks.

1 Introduction

Pattern Recognition (PR) has numerous well-established sub-areas such as statistical, syntactic, structural and neural. The field of *Chaotic* PR is, however, relatively new and is founded on the principles of chaos theory. It is also based on a distinct phenomenon, namely that of *switching* from *chaos* to *periodicity*. Indeed, Freeman's clinical work has clearly demonstrated that the brain, at the individual neural level and at the global level, possesses chaotic properties. He showed that the quiescent state of the brain is chaos. However, during perception, when attention is focused on any sensory stimulus, the brain activity becomes more periodic [7].

If the brain is capable of displaying both chaotic and periodic behavior, the premise of this paper is that it is expedient to devise artificial Neural Network (NN) systems that can display these properties too. Thus, the primary goal of *chaotic* PR is to develop a system which mimics the brain to achieve chaos and PR, and to consequently develop a new PR paradigm in itself.

Philosophically, the question of whether we can achieve *chaotic* PR by using non-dynamic methods seems to have an unequivocal negative response. The subsequent question of whether we can accomplish the goal of designing a *chaotic* PR system with a *single* computing element (or neuron) also seems improbable. It appears as if we have to resort to a *network* of neurons, and any such network which possesses *chaotic* properties is referred to as a Chaotic Neural Network (CNN). The fundamental issue, really, is whether we can drive such a NN from chaos to periodicity and *vice versa* by merely controlling the input patterns.

At a very fundamental level, the field of NNs deals with understanding the brain as an information processing machine. Conversely, from a computational perspective, it concerns utilizing the knowledge of the (human) brain to build more intelligent computational *models* and computer systems. Thus, NNs have been widely and successfully applied to an ensemble of information processing problems, and the areas of PR and forecasting are rather primary application domains.

Before we proceed, it is fitting for us to submit a consistent terminology, i.e., to distinguish between systems that possess Associative Memory (AM) and

Pattern Recognition (PR). The reader should note that our primary concern is the latter. The literature reports that AM has two forms: auto-association and hetero-association. In auto-association, a NN is required to store a set of patterns. Whenever a single pattern or its "distorted" version is presented to the NN, a system possessing auto-association should be able to recall the particular pattern correctly. Hetero-association differs from auto-association in that an arbitrary set of input patterns is paired with another arbitrary set of output patterns. As opposed to these, the phenomenon of PR is mainly concerned with "classification".

CNNs which also possessed a *weak* form of PR were first proposed by Adachi and his co-authors [1–5]. It would be fair to state (and give them the honor) that they pioneered this "new" field of CNNs. In their papers, by modifying the discrete-time neuron model of Caianiello, Nagumo and Sato, they designed a novel and simple neural model possessing chaotic dynamics, and an Artificial Neural Network (ANN) composed of such chaotic neurons. Their experimental results demonstrated that such a CNN (referred to the AdNN in this paper) possesses both AM and PR properties, as will be illustrated in Section 2.

Historically, the initial and pioneering results concerning these CNNs were presented in [1–5]. In the next year, the author of [8] proposed two methods of controlling chaos by introducing a small perturbation in continuous time, i.e., by invoking a combined feedback with the use of a specially-designed external oscillator or by a delayed self-controlling feedback without the use of any external force. The reason for the introduction of this perturbation was to stabilize the unstable periodic orbit of the chaotic system. Subsequently, motivated by the work of Adachi, Aihara and Pyragas, various types of CNNs have been proposed to solve a number of optimization problems (such as the Traveling Salesman Problem, (TSP)), or to obtain AM and/or PR properties. Later, the author of [9] proposed a NN model based on the Globally Coupled Map (GCM) derived by modifying the Hopfield network. It was reported that this model was capable of information processing and solving the TSP. Chen and Aihara reported a new method referred to as chaotic simulated annealing based on a NN model with transient chaos properties [10]. They also numerically investigated the transiently chaotic neurodynamics with examples of a single neuron model and the TSP. Their conclusion was that this new model possessed a higher searching ability for solving combinatorial optimization problems when compared to both the Hopfield-Tank approach and a stochastic simulated annealing scheme. An interesting step in this regard was the work reported in [11], where the authors utilized the delayed feedback and the Ikeda map to design a CNN to mimic the biological phenomena observed by Freeman [7].

From a different perspective, in [12–14], Hiura and Tanaka investigated several CNNs based on a piecewise sine map or the Duffing's equation to solve the TSP. Their results showed that the latter yielded a better performance than the former.

More recently, based on the AdNN, Calitoiu and his co-authors made some interesting modifications to the basic network connections so as to obtain PR properties and "blurring". In [15], they showed that by binding the state variables to those associated with *certain* states, one could obtain PR phenomena. However, by modifying the manner in which the state variables were bound, they designed a newly-created machine, the so-called Mb-AdNN, which was also capable of justifying "blurring" from a NN perspective.

While all of the above are both novel and interesting, since most of these CNNs are *completely*-connected graphs, the computational burden is rather intensive. Aiming to reduce the computational cost, in our previous paper [16], we proposed a mechanism (the Linearized AdNN (L-AdNN)) to reduce the computational load of the AdNN. This was achieved by using a spanning tree of the complete graph, and invoking a gradient search algorithm to compute the edge weights, thus minimizing the computation time to be linear.

Although it was initially claimed that the AdNN and M-AdNN possessed "pure" (i.e., periodic) PR properties, in actuality, this claim is not as precise as the authors claimed. We shall now clarify this. As explained in detail in [16, 17], if the AdNN weights the external inputs with the quantities determined by $a_i = 2 + 6x_i$, the properties of the system become closer to mimicing a PR phenomenon. Adachi *et al.* claimed that such a NN yields the stored pattern at the output periodically, with a short transient phase and a small periodicity. However, it turns out that if *untrained* patterns are presented to the system, they also appear periodically, although with a longer periodicity. Besides, the untrained input patterns can lead to having the system occasionally output *trained* patterns. Although the former can be considered to be a PR phenomenon, the latter is a handicap because we would rather prefer the system to be chaotic for all untrained patterns - which the AdNN cannot achieve for these settings. This phenomenon is also pertinent for the M-AdNN, i.e., the output can be periodic for both trained and untrained inputs.

To complete this historical overview, we mention that in [16], we showed that the AdNN goes through a spectrum of characteristics (i.e., AM, *quasi*-chaotic, and PR) as one of its crucial parameters, α, changes. In particular, it is even capable of recognizing masked or occluded patterns.

The primary aim of this paper is to show that the M-AdNN, when tuned appropriately, is capable of demonstrating ideal PR capabilities. Indeed, we shall argue that this NN, referred to as the Ideal-M-AdNN:

- Can be trained by a set of patterns. This will determine the weights of the CNN.
- When the Ideal-M-AdNN is presented with a trained pattern (or its noisy version) at the input, it resonates and yields *only this* pattern at the output periodically. We emphasize that, in this case, no other trained pattern is observed at the output.
- When the Ideal-M-AdNN is presented with an untrained pattern at the input, it continues to behave chaotically. In this case, none of the trained patterns is observed at the output - thus yielding the gold standard of *chaotic* PR.

One should remember that the method for setting $\{w_{i,j}\}$ and $\{a_i\}$ were already specified for the AdNN in the paper by Adachi and Aihara. But the reader should remember that we are not working with the AdNN, but rather with the M-AdNN, in which the state bindings are significantly different. Apart from making the changes in the structure, the authors of the M-AdNN also made changes in the assignment strategies. Our task in this paper is: (a) To demonstrate that these latter changes are not warranted; (b) To argue that in the paper due to Adachi and Aihara, the choice of $a = 2$ is not the best value that can yield AM properties; and (c) To demonstrate that the value of one of its parameters, ε (explained presently) that was used, is just a *single* one of the choices that can force the system to be chaotic.Thus, **the primary contributions of the paper** are:

1. We formalize the requirements of a PR system which is founded on the theory of chaotic NNs;
2. We enhance the M-AdNN to yield the Ideal-M-AdNN, so that it does, indeed, possess *Chaotic* PR properties;
3. We show that the Ideal-M-AdNN does switch from chaos to periodicity when it encounters a trained pattern, but that it is truly chaotic for all other input patterns;
4. We present results concerning the stability of the network and its transient and dynamic retrieval characteristics. This analysis is achieved using eigenvalue considerations, and the Lyapunov exponents;
5. We provide explicit experimental results justifying the claims that have been made.

We conclude by mentioning that in this paper, we were not so concerned about the application domain. Rather, our work is intended to be of an investigatory sort.

2   State of the Art

The AdNN and Its Variants: The AdNN is a network of neurons with weights associated with the edges, a well-defined Present-State/Next-State function, and a well-defined State/Output function. It is composed of N neurons which are topologically arranged as a completely connected graph. A neuron, identified by the index i, is characterized by two internal states, $\eta_i(t)$ and $\xi_i(t)$ $(i = 1...N)$ at time t, whose values are defined by Equations (1) and (2) respectively. The output of the ith neuron, $x_i(t)$, is given by Equation (3), which specifies the so-called *Logistic Function*.

$$\eta_i(t+1) = k_f\eta_i(t) + \sum_{j=1}^{N} w_{ij}x_j(t), \tag{1}$$

$$\xi_i(t+1) = k_r\xi_i(t) - \alpha x_i(t) + a_i. \tag{2}$$

$$x_i(t+1) = f(\eta_i(t+1) + \xi_i(t+1)). \tag{3}$$

where $f(\cdot)$ is defined by $f(x) = 1/(1 + e^{-x})$. Under certain settings, the AdNN can behave as a dynamic AM. It can dynamically recall all the memorized patterns as a consequence of an input which serves as a "trigger". If the external stimulations correspond to trained patterns, the AdNN can also behave like a PR system, although with some limitations and weaknesses, as will be explained below, and as also illustrated in [18].

By invoking a Lyapunov Exponents (LE) analysis, one can show that the AdNN has $2N$ *negative* LEs. In order to obtain a system with *positive* LEs, Calitoiu *et al* [15] proposed a model of CNNs, called the M-AdNN, which modifies the AdNN to enhance its PR capabilities.

The fundamental difference between the AdNN and the M-AdNN in terms of their Present-State/Next-State equations is that the latter has only a *single* global neuron (and its corresponding two global states) which is used for the state updating criterion for *all* the neurons. Thus, $\eta_i(t)$ and $\xi_i(t)$ are updated as per:

$$\eta_i(t+1) = k_f\eta_m(t) + \sum_{j=1}^{N} w_{ij}x_j(t) \tag{4}$$

$$\xi_i(t+1) = k_r\xi_m(t) - \alpha x_i(t) + a_i. \tag{5}$$

$$x_i(t+1) = f(\eta_i(t+1) + \xi_i(t+1)). \tag{6}$$

Observe that at time $t+1$, the global states are updated with the values of the states of *the single* neuron, $\eta_m(t)$ and $\xi_m(t)$. The value of m is set to be N. This is in contrast to the AdNN in which the updating at time $t+1$ uses the internal state values of *all* the neurons at time t. The weight assignment rule for the M-AdNN is the classical variant $w_{ij} = \frac{1}{p}\sum_{s=1}^{p} x_i^s x_j^s$. This is in contrast to the AdNN which uses $w_{ij} = \frac{1}{p}\sum_{s=1}^{p}(2x_i^s - 1)(2x_j^s - 1)$. Additionally, Calitoiu *et al.* set $a_i = x_i^s$, as opposed to the AdNN in which $a_i = 2 + 6x_i^s$. Thus, as the authors demonstrated, the M-AdNN will be more "receptive" to external inputs, leading to a chaotic PR system. However, we will illustrate that the parameters Calitoiu *et al.* set were not the appropriate ones to yield PR, and show how these settings must be tuned to yield a more superior PR system. This will be done in Sections 3.1 and 3.2.

We list below the relevant salient features of the M-AdNN in the interest of completeness and continuity:

1. Being a variant of the AdNN, the M-AdNN is topologically a completely-connected NN;
2. The main difference between the M-AdNN and the classical auto-AM model is that unlike the latter, the M-AdNN was seen to be chaotic, this behavior being a consequence of the dynamics of the underlying system. The M-AdNN seems to be more representative of the way by which the brain achieves PR;

3. The most significant difference between the AdNN and the M-AdNN is the way by which the values of the internal state(s) are updated. The M-AdNN uses two global internal states, both of which are associated with a *single* neuron. By using these global states, the transient phase is reduced to be approximately 30 [6], which is in contrast to the AdNN's transient phase – which can be as large 21,000;

4. The authors of [6] show that the largest Lyapunov exponent of the AdNN is $\lambda = \log k_r < 0$ when $k_r = 0.9$, which implies that AdNN is not really chaotic. In contrast, the M-AdNN has a positive largest Lyapunov exponent given by $\lambda = \frac{1}{2} \log N + \log k_r$, rendering it chaotic.

5. The experimental results reported in [6] claimed that the M-AdNN responds periodically for a trained input pattern. However, it turns out that this periodic output occurs for both trained and untrained patterns, as can be seen in [16].

3   The Ideal-M-AdNN

Traditionally, PR systems work with the following model: Given a set of training patterns, a PR system learns the characteristics of the class of the patterns, and retains this information either parametrically or non-parametrically. When a testing sample is presented to the system, a decision of the identity of the sample class is made using the corresponding "discriminant" function, and this class is "proclaimed" by the system as the identity of the pattern.

The goal of the field of Chaotic PR systems can be expressed as follows: We do not intend a chaotic PR system to report the identity of a testing pattern with such a "proclamation". Rather, what we want to achieve for a the chaotic PR system are the following phenomena:

– The system must yield a strong *periodic* signal when a trained pattern is to be recognized.
– Further, between two consecutive recognized patterns, none of the trained patterns must be recalled.
– On the other hand, and most importantly, if an untrained pattern is presented, the system must be chaotic.

Calitoiu *et al* were the first researchers who recorded the potential of chaotic NNs to achieve PR. But unfortunately, their model, as presented in [6], named the M-AdNN, was not capable of demonstrating *all* the PR properties mentioned above. To be more accurate:

– For trained patterns, the M-AdNN reacts exactly as what we want: The specific input patterns appear at the output *periodically*.
– However, for untrained patterns, the M-AdNN still yields the unexpected *untrained* pattern periodically.

– In summary, the M-AdNN yields at the output whatever is presented to the system as its input[1], and therefore we claim that the M-AdNN is not good enough for PR since it is not able to resonate *only* for the trained patterns.

Nevertheless, the M-AdNN is still a fascinating network since it can produce periodic output while it is still in the midst of chaotic iterations, which, since it was previously unreported in the literature, is really a ground-breaking work in the area of chaotic NNs. Based on the results of Calitoiu *et al*, we improve the M-AdNN for it to truly possess *Chaotic* PR properties. The new model proposed here is referred as the "Ideal-M-AdNN".

The topology of the Ideal-M-AdNN is exactly the same as that of the M-AdNN, which is, on deeper consideration, seen to actually also be a Hopfield-like model. Structurally, it is also composed of N neurons, topologically arranged as a completely connected graph. Again, each neuron has two internal states $\eta_i(t)$ and $\xi_i(t)$, and an output $x_i(t)$. Just like the M-AdNN, the Present-State/Next-State equations of the Ideal-M-AdNN are defined in terms of only a *single* global neuron (and its corresponding two global states), which, in turn, is used for the state updating criterion for *all* the neurons. Thus, $\eta(t)$ and $\xi(t)$ are updated as per:

$$\eta_i(t+1) = k_f \eta_m(t) + \sum_{j=1}^{N} w_{ij} x_j(t), \tag{7}$$

$$\xi_i(t+1) = k_r \xi_m(t) - \alpha x_i(t) + a_i, \tag{8}$$

$$x_i(t+1) = f(\eta_i(t+1) + \xi_i(t+1)), \tag{9}$$

where m is the index of this so-called "global" neuron.

We shall now concentrate on the differences between the two models, which are the parameters: $\{w_{ij}\}$, ε and a_i. In the work reported in [6], these parameters were arbitrarily set to be $w_{ij} = \frac{1}{4}\Sigma_{s=1}^{4} x_i^s x_j^s$, $\varepsilon = 0.00015$ and $a_i = x_i$. In this paper, we will show that these values must not be set arbitrarily. Rather, we shall address the issue of how these parameters must be assigned their respective values so as to yield all the properties required of a *Chaotic* PR system.

3.1   The Weights of the Ideal-M-AdNN

In the interest of clarity, we first argue that the weights assignment Equation (10) for the M-AdNN is not appropriate. We rather advocate another form of the Hebbian rule given by Equation (15) instead. The reason is explained as below: As we have observed above, the M-AdNN uses a form of the Hebbian rule to determine the weights of the connections in the network. This rule is defined by the following equation:

---

[1] This phenomenon was not observed nor reported in [6].

$$w_{ij} = \frac{1}{p} \sum_{s=1}^{p} P_i^s P_j^s, \tag{10}$$

where $\{P\}$ are the training patterns, P_i^s denotes the ith neuron of the sth pattern P^s, and p is the number of training patterns.

At this juncture, we emphasize that the Hebbian rule, Equation (10), is founded on one fundamental premise: Any pair of learning vectors, P and Q, must be orthogonal, and thus, for all P and Q, if P^T is the transpose of P:

$$P^T Q = \begin{cases} 0, (P \neq Q) \\ N, (P = Q). \end{cases} \tag{11}$$

The reasons why the formula given by Equation (10) is based on the above, is because of the following:

If P denotes an N-by-1 vector (e.g, the $P1$ of Figure 8 (a) is a 100-by-1 vector), Equation (10) can be rewritten as:

$$W = \frac{1}{p} \sum_{s=1}^{p} P^s (P^s)^T. \tag{12}$$

Thus, the output for any given input vector P^k is:

$$O = W P^k = \frac{1}{p} \sum_{s=1}^{p} P^s (P^s)^T P^k$$

$$= \frac{1}{p} \sum_{s=1}^{p} P^s \left[(P^s)^T P^k \right] = \frac{N}{p} P^k \tag{13}$$

As per Equation (11), the output $O = \frac{N}{p} P^k$ is a scalar[2] of the input P^k. Otherwise, if the learning vectors $\{P\}$ are not orthogonal, Equation (13) has the form:

$$O = \frac{1}{p} \sum_{s=1}^{p} P^s [(P^s)^T P^k] = \frac{1}{p} \left(N P^k + P^{noise} \right), \tag{14}$$

where $P^{noise} = \sum_{s=1, s \neq k}^{p} b_s P^s$ and $b_s = (P^s)^T P^k$. Obviously, $O = \frac{N}{p} P^k$ if and only if $P^{noise} = 0$, explaining the rationale for orthogonality.

The need for orthogonality is not so stringent, because, as per the recorded research, when the number of neurons is much larger than the number of patterns, and the learning vectors are randomly chosen from a large set, the probability of having the learning vectors to be orthogonal is very high. Consequently, Equation (11) is true, albeit probabilistically.

Summary: Based on the above observations, we conclude that for the M-AdNN, we should not use the Hebbian rule as dictated by the form given in Equation

---

[2] Observe that without loss of generality, we can easily force $O = P^k$ instead of $O = \frac{N}{p} P^k$, by virtue of a straightforward normalization.

(10), since the data sets used by both Adachi *et al* and Calitoiu *et al* are defined on $\{0,1\}^N$, and the output is further restricted to be in $[0,1]$ by virtue of the logistic function. We may easily verify that any pair of the given patterns in Figure 8 are NOT orthogonal. In fact, this is why Adachi and Aihara computed the weight by scaling all the patterns to -1 and 1 using the formula given by Equation (15) instead of (10):

$$w_{ij} = \frac{1}{p} \sum_{s=1}^{p} (2P_i^s - 1)(2P_j^s - 1). \tag{15}$$

By virtue of this argument, in this paper, we advocate the use of Equation (15), to determine the network's weights.

It is pertinent to mention that since the patterns P are scaled to be in the range between -1 and 1, it *does* change the corresponding property of orthogonality. Actually, it is very interesting to compare the inner product of the so-called Adachi dataset before and after scaling (please see Table 1). From it we see that the inner products of any pair of *scaled* patterns while not being exactly 0, are very close to 0 – i.e., these patterns are *almost* orthogonal. In contrast, the inner products of the corresponding *unscaled* patterns are very large. Again, this table confirms that Equation (15) is a more suitable (and reasonable) expression than Equation (10).

Table 1. The inner products of the patterns of the Adachi set. In the left table, all patterns are scaled to -1 and 1 by $2P - 1$. In the table on the right, all the patterns are defined on $\{0,1\}^N$. The inner products in the table on the left are much smaller and almost 0, implying that the patterns are *almost* orthogonal.

	P1	P2	P3	P4		P1	P2	P3	P4
P1	100	10	10	6	P1	49	27	27	26
P2	10	100	-4	4	P2	27	50	24	26
P3	10	-4	100	12	P3	27	24	50	28
P4	6	4	12	100	P4	26	26	28	50

3.2 The Steepness, Refractory and Stimulus Parameters

Significance of ε for the AdNN. The next issue that we need to consider concerns the value of the steepness parameter ε of the output function. From Section 2 we see that the output function is defined by the Logistic function $f(x) = \frac{1}{1+e^{-x/\varepsilon}}$, which is a typical sigmoid function.

One can see that ε controls the steepness of the output. If $\varepsilon = 0.01$, then $f(x)$ is a normal sigmoid function. If ε is too small, say, 0.0001, the Logistic function almost degrades to become a unit step function (see Figure 1). The question, really, is one of knowing how to set the "optimal" value for ε. To provide a rationale for determining the best value of ε, we concentrate on the Adachi's neural model [2] defined by:

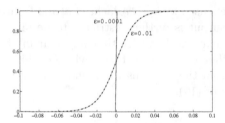

Fig. 1. A graph demonstrating the effect of the parameter ε to change the steepness of the sigmoidal function. The ε for the dash and solid lines are 0.01 and 0.0001 respectively.

$$y(t+1) = ky(t) - \alpha f(y(t)) + a, \tag{16}$$

where $f(\cdot)$ is a continuous differentiable function, which as per Adachi *et al* [2], is the Logistic function[3].

The properties of Equation (16) greatly depend on the parameters k, α, a and $f(\cdot)$. In order to obtain the full spectrum of the properties represented by Equation (16), it is beneficial for us to first consider $f(\cdot)$ in terms of a unit step function, and to work with a fixed point analysis. To do this, we consider a few distinct cases listed below:

1. If the system has only *one* fixed point.

- In this case, obviously, $y(t) = 0$ is not a fixed point of Equation (16) unless $a = 0$.
- If Equation (16) has only one positive fixed point, say y_0, then the value of this point can be resolved from Equation (16) as $y_0 = \frac{a-\alpha}{1-k}$. In the paper by Nagumo and Sato [19], the authors set $\alpha = 1$, $k < 1$, and let a vary from 0 to 1. Apparently, the above value of y_0 is always negative, which contradicts our assumption that y_0 is a positive fixed point. Similarly, we can see that Equation (16) does not have any negative fixed points either. In other words, we conclude that Equation (16) does not have any fixed point.

2. If the system has period-2 points, say y_1 and y_2 (where $y_1 \neq y_2$), we see that these points must satisfy:

$$\begin{cases} y_2 = ky_1 - \alpha f(y_1) + a \\ y_1 = ky_2 - \alpha f(y_2) + a. \end{cases} \tag{17}$$

- If 0 is one of the period-2 points (without loss of generality, $y_1 = 0, y_2 \neq 0$), then we can solve the above equations to yield the two solutions as:

---

[3] Historically, the original form of this equation initially appeared in the paper by Nagumo and Sato [19]. The only difference between the function that these researchers used and the one discussed here is the function $f(\cdot)$, since in [19], they utilized the unit step function instead of the logistic function.

$$\begin{cases} y_2 = a \\ y_1 = (k+1)a - \alpha. \end{cases}$$

From this, we see that the two points exist if and only if $a \neq 0$ and $\alpha = (1+k)a$.

- If the system has two positive period-2 points (assume $y_1 > 0$, $y_2 > 0$), we see that the system can be rewritten as:

$$\begin{cases} y_2 = ky_1 - \alpha + a \\ y_1 = ky_2 - \alpha + a, \end{cases}$$

which is equivalent to implying that $y_2 - y_1 = k(y_1 - y_2)$. Clearly, this equality cannot hold since $y_1 \neq y_2$ and $k > 0$. In an analogous manner we can conclude that the system does not have two negative period-2 points either.

- If the system has two period-2 points with different signs (without loss of generality we can assume that $y_1 > 0$, $y_2 < 0$), we solve Equation (17) as:

$$\begin{cases} y_1 = \frac{a(k+1)-k\alpha}{1-k^2} \\ y_2 = \frac{a(k+1)-\alpha}{1-k^2}. \end{cases} \tag{18}$$

Since $y_1 > 0$ and $y_2 < 0$, this yields the inequality:

$$\frac{k\alpha}{1+k} < a < \frac{\alpha}{1+k}. \tag{19}$$

In Nagumo's paper [19], the authors set $k = 0.5$ and $\alpha = 1$, implying that the system has period-2 fixed points only when $1/3 < a < 2/3$.

From the above, we know the Nagumo-Sato model does not have any fixed points, but that it rather does have period-2 points. The next question that we have to resolve is whether it possesses chaotic properties? To determine this, we have to investigate what happens to the iteration orbits when the initial point is not exactly on, but close to a period-2 point? Indeed, this question can be answered by considering the inequality:

$$\left| \frac{\partial g^{(2)}(y)}{\partial y} \right|_{y=y_1} < 1, \tag{20}$$

where $g(y) = ky - \alpha f(y) + a$, and $f(\cdot)$ is still a unit step function. A simple algebraic manipulation displays that the inequality given by Equation (20) is always true whenever $k < 1$, implying that whatever the initial condition, after a sufficiently large number of iterations, the orbits converge to the period-2 points. In other words, all the period-2 points are attracting, implying that the system does not demonstrate any chaotic phenomena.

3. If we now consider the period-n orbits, one can see that it is not possible to resolve the set of equations that generalize from Equations (18) and (19) for n variables. However, the analysis that corresponds to generalizing the *stability*

equation is exactly the same as in Equation (20), and in every case, one can again show that the inequality is true[4]. This again leads to the conclusion that all the period-n orbits are also attracting. This can be verified by considering the graphs in Figure 2.

The situation is, however, completely different if the unit step function is replaced by the sigmoid function. This is a relevant issue because, as stated earlier, the difference between the Nagumo-Sato model and the Adachi model lies in the output function $f(\cdot)$, where the former uses the unit step function and the latter uses a sigmoid function. The issue is further accentuated because the logistic function may degrade to an unit step function if an inappropriate value of ε is utilized, as shown in Figure 1. Indeed, when ε is very small, the sigmoid function will "degrade" to the unit step function, leading us to the analysis which we have just concluded.

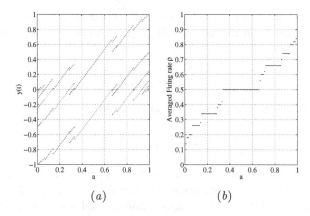

(a) $\qquad\qquad\qquad\qquad$ (b)

Fig. 2. The period-n orbits (a) and the averaged firing rate (b) of the Nagumo-Sato model, in which a varies from 0 to 1.

Significance of ε for the M-AdNN. Since the M-AdNN is a modified version of the AdNN (which uses the Logistic function as the output function), setting the value of ε is also quite pertinent in our case. As we shall argue now, in order to enable the system to have chaotic properties, the parameter ε should not be close to zero.

To demonstrate the role of parameter ε of the Adachi model, as before, we analyze the behavior of the model starting from fixed points or period-n points.

Let us first analyze whether the Adachi neuron has any fixed points, i.e., if

$$y(t+1) = y(t) = ky(t) - \alpha f(y(t)) + a$$

---

[4] This has been done for values of n which equal 3, 4 and 5. The general analysis for arbitrary values of n can also be done using the Schwarz derivative and the period doubling bifurcation theorem. However, this goes beyond the scope of this paper and is thus omitted here.

is ever satisfied. Since this equation is a transcendental equation, obtaining the exact fixed point(s) is far from trivial. Nevertheless, we are still able to achieve some analysis on the bifurcation parameters.

As we know, if a fixed point(s) y^* exists, it should satisfy:

$$\alpha f(y^*) = ky^* - y^* + a.$$

Our first task is to see if a fixed point y^* does exist.

To do this, we compose:

$$F(y) = -\alpha f(y) + (k - 1)y + a, \tag{21}$$

which is done so that the root of $F(y)$ corresponds to the fixed point of y.

It is easy to observe that following: For any given negative y, $f(y)$ is almost zero by virtue of its sigmoidal nature. Thus, $F(y) \approx (k - 1)y + a > 0$. Similarly, for any given positive y, (for example, $y=1$), $f(y)$ is almost unity, again by virtue of its sigmoidal nature. Thus, $F(y) \approx -\alpha + k - 1 + a < 0$. Since $F(y)$ is a continuous function over $(-\infty, \infty)$, and since it changes its sign, there must exist at least one y^* satisfying $F(y^*) = 0$ implying that this value of y^* is the fixed point.

The next question is: How many fixed points does the system have? This question can be answered by verifying the monotonicity of Equation (21). Since $f'(y) = -\frac{1}{\varepsilon}f(y)[1 - f(y)]$, we get:

$$F'(y) = -\frac{\alpha}{\varepsilon}f(y)[1 - f(y)] + k - 1. \tag{22}$$

Again, since we always have $f(y)[1 - f(y)] \approx 0$, it implies that $F'(y) \approx k - 1 < 0$, which, in turn means that $F(y)$ is a decreasing function. As a result of the fact that it is positive on one side of the fixed point, negative on the other side, and simultaneously always has a negative derivative, we conclude that the Adachi model has only a *single* fixed point, y^*.

If y^* is stable, it should satisfy the condition

$$\left|\frac{dy(t + 1)}{dy(t)}\right|_{y=y^*} < 1. \tag{23}$$

From the dynamical form of $y(t + 1)$, this derivative can be rewritten as

$$\left|k - \frac{\alpha}{\varepsilon}f(y^*)(1 - f(y^*))\right| < 1, \tag{24}$$

which is equivalent to:

$$\begin{cases} \alpha f^2(y) - \alpha f(y) + \varepsilon(k + 1) > 0 \\ \alpha f^2(y) - \alpha f(y) + \varepsilon(k - 1) < 0. \end{cases} \tag{25}$$

Since each neuron's output is between 0 and 1, we see that for all values in $(0,1)$ the inequalities given by Equation (25) are always true. This leads us to the discriminants:

$$\begin{cases} \Delta_1 = \alpha^2 - 4\alpha\varepsilon(k + 1) < 0 \\ \Delta_2 = \alpha^2 - 4\alpha\varepsilon(k - 1) > 0. \end{cases} \tag{26}$$

Obviously, since $k - 1 < 0$, $\alpha > 0$ and $\varepsilon > 0$, the second inequality is always true. Concentrating now on the first inequality, we can solve for ε and see that it has to satisfy:

$$\varepsilon > \frac{\alpha}{4(k+1)}. \tag{27}$$

Briefly, this states that if the fixed point is stable, ε should be greater than $\alpha/4(k+1)$; otherwise, it is not stable.

This condition is experimentally verified in Figures 3 (a), (b) and (c). In our experiment, the parameters were set to be $\alpha = 1, k = 0.5$. In this case, the "tipping point" for ε is $1/6 \approx 0.1667$. As one can very clearly see, if $\varepsilon = 0.18 > 0.1667$, all of the fixed points are stable (Figure 3 (a)); Otherwise, if $\varepsilon = 0.15 < 0.1667$, there exist period doubling bifurcations[5] (Figure 3 (b)). As ε is further decreased, one can observe chaotic windows (Figure 3 (c)).

We conclude this section by emphasizing that ε cannot be too small, for if it were, the Adachi neural model would degrade to the Nagumo-Santo model, which does not demonstrate any chaotic behavior. This is also seen from Figures 4 (a) and (b). In (a), where the parameter ε is 0.015, we can see that the orbit of the iteration is non-periodic, while in (b), in which ε is 0.00015, the orbit of the trajectory is forced to have a periodicity of 5. In fact, we have calculated the whole LE spectrum of the system (16) when it concerns the parameters ε, as shown in Figure 5 (a). In order to get a better view, we have amplified the graph in the area $\varepsilon \in [0, 0.04]$ and in the area $\varepsilon \in [-1.5, 0.5]$ in (b). As one can see when $\varepsilon = 0.015$, the LE is positive, and has a value which is approximately 0.3, which indicates chaos. On the other hand, when $\varepsilon = 0.00015$, the LE is negative. Of course, there are some other optional values which also lead to chaos, such as $\varepsilon = 0.007, 0.008, 0.012$ etc can also lead to chaos.

Indeed, our arguments show that the value of ε as set in [15] to be $\varepsilon = 0.00015$, is not appropriate. Rather, to develop the Ideal-M-AdNN, we still opted to use a value of ε which is *two* orders of magnitude larger, i.e., $\varepsilon = 0.015$.

4 Lyapunov Exponents Analysis of the Ideal-M-AdNN

We shall now analyze the Ideal-M-AdNN, both from the perspective of a *single* neuron and of the entire network.

As is well known, the LEs describe the behavior of a dynamical system. There are many ways, both numerically and analytically, to compute the LEs of a dynamical system. Generally, for systems with a small dimension, the best way is to analytically compute it using its formal definition. As opposed to this, for systems with a high dimension, it is usually not easy to obtain the entire LE spectrum in an analytic manner. In this case, we have several other alternatives to judge whether a system is chaotic. One of these is to merely determine the largest LE (instead of computing the entire LE spectrum) since the existence

---

[5] In this paper, we analyze the bifurcation in a rather simple way. The reader must note that this could also be achieved by using the Schwarz derivative and the period doubling bifurcation theorem.

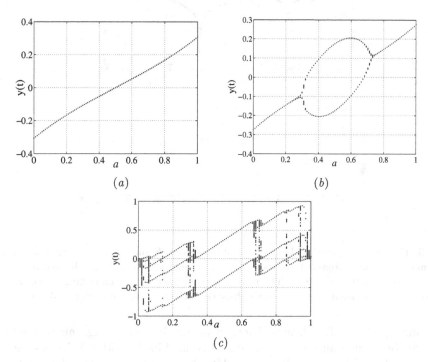

Fig. 3. This figure shows that as the steepness parameter, ε, varies, the dynamics of the Adachi neuron changes significantly. In the three figures, the values of ε are 0.18, 0.15 and 0.015 respectively.

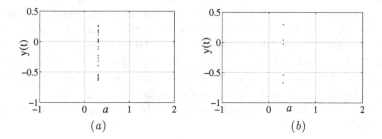

Fig. 4. This figure shows that if the value of ε is too small, it will force the orbit of the trajectory to be periodic (b). The values of ε for (a) and (b) are 0.015 and 0.00015 respectively.

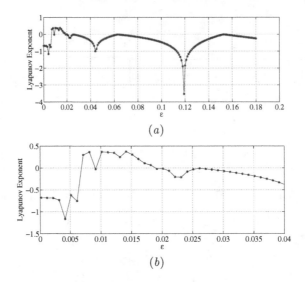

Fig. 5. This figure shows the LE spectrum of the system. Some values between 0.007 and 0.02 may cause the system to be chaotic. The reader should observe that other parametric settings are: $k = 0.5$, $a = 0.31$. The reason why we have used such settings is because one can observe chaotic windows under these conditions, as shown in Figure 3 (c).

of a single positive LE indicates chaotic behavior. In this setting, this is usually achieved by using a numerical scheme as described in Algorithm 1. In this algorithm, the basic idea is to follow two orbits that are close to each other, and to calculate their average logarithmic rate of separation [20–22].

In practice, this algorithm is both simple and convenient if we have the right to access the equations that govern the system. Furthermore, if it is easy to obtain the partial derivatives of the system, we can also calculate the LE spectrum by QR decomposition [20, 23, 22].

We now present the theoretical basis for obtaining the LE spectrum using the QR decomposition. To do this, in the interest of compactness, we use the notation that J_t represents the Jacobian matrix at time t, i.e., $J_t = J(t)$, whose transpose is written as J_t^T. Also, x_0 is assumed to be a randomly chosen starting point, and $\{f(x_0), f_2(x_0), f_3(x_0) \cdots f_t(x_0)\}$ denotes the trajectory obtained as a consequence of iterating as per the system's dynamical equation. Again, to simplify the notation, we shall use the notation that $x_k = f_k(x_0)$.

As we know, the LEs of a network are defined by the eigenvalues of the limit of the matrix[6]

$$\Lambda = \lim_{t \to \infty} [J_t \cdot J_t^T]^{\frac{1}{2t}}. \tag{28}$$

Generally, it is hard to compute the limit of Equation (28), which is why one often invokes the QR decomposition. To achieve this, we note that by the chain rule,

---

[6] The conditions for the existence of the limit are given by the Oseledec theorem.

Algorithm 1. Numerical Calculation of the Lyapunov Exponent

1: Start with any initial condition in the basin of attraction.
2: Iterate until the orbit is on the attractor.
3: Select a nearby point (separated by d_0, as indicated in [21]. In our work, we set $d_0 = 10^{-12}$).
4: Advance both orbits one iteration and calculate the new separation d_1.
5: Evaluate $\log |d_1/d_0|$ in any convenient base.
6: Readjust one orbit so that its separation is d_0 in the same direction as d_1.
7: Repeat Steps 4-6 many times and calculate the average of Step 5.

$$Df_t(x_0) = J(f_{t-1}(x_0)) \cdots J(f(x_0)) \cdot J(x_0),$$

where $J(x_i) = Df(x)|_{x=x_i} = J(f_i(x_0))$. As clarified in [20, 22], we can rewrite $J_0(x_0)$ using its QR decomposition as $J_0(x_0) = Q_1 R_1$, where Q_1 is orthogonal implying that $Q_1 \cdot Q_1^T = I$. If we define $J_k^* = J(f_{k-1}(x_0))Q_{k-1}$, we can write *its* QR decomposition as $J_k^* = Q_k R_k$. Thus, $J(f_{k-1}(x_0)) \cdot Q_{k-1} = Q_k R_k$. Consequently, we obtain $J(f_{k-1}(x_0)) = Q_k R_k Q_{k-1}^{-1}$. By applying this equation to the chain rule, the differential $Df_t(x_0)$ can be transformed to be:

$$Df_t(x_0) = Q_t R_t Q_{t-1}^{-1} \cdot Q_{t-1} R_{t-1} Q_{t-2}^{-1} \cdots R_1$$
$$= Q_t R_t \cdots R_1. \tag{29}$$

The LEs, $\{\lambda_i\}$, can then be obtained as:

$$\lim_{t\to\infty} \frac{1}{t} \ln |v_{ii}(t)| = \lambda_i, \tag{30}$$

where $\{v_{ii}(t)\}$ are the diagonal elements of the product $R_{t-1}R_{t-2}\cdots R_1$.

Again, for the reader's convenience, the algorithmic details are formally given in Algorithm 2.

4.1 Numeric Lyapunov Analysis of a Ideal-M-AdNN

We shall now undertake a Lyapunov Analysis of the Ideal-M-AdNN. To do this, we first undertake the Lyapunov analysis of a single neuron in the interest of simplicity. Indeed, it can be easily proven that a single neuron is chaotic when the parameters are properly set.

Consider a primitive component of the Ideal-M-AdNN, where the model of a *single* neuron can be described as[7]:

$$\eta(t+1) = k_f \eta(t) + wx(t), \tag{31}$$
$$\xi(t+1) = k_r \xi(t) - \alpha x(t) + a, \tag{32}$$
$$x(t+1) = \frac{1}{1 + e^{-(\eta(t+1)+\xi(t+1))/\varepsilon}}. \tag{33}$$

---

[7] It should be observed that $w_{ii} = 1, i = 1, 2, \cdots, N$ for the entire network. For a single neuron, the value of w should be $w = 1$. We should observe that, for a primitive neuron, there is no difference between the AdNN, the M-AdNN and the Ideal-M-AdNN.

Algorithm 2. Calculation of the Lyapunov Exponent by QR Decomposition

---

1: Choose a starting point x_0 and compute the Jacobian matrix $J_0(x_0) = \frac{df(x)}{dx}$ at x_0.
2: Decompose $J_0(x_0)$ by the QR decomposition: $J_0(x_0) = Q_1 \cdot R_1$ where Q_1 is an orthogonal matrix and R_1 is an upper triangular matrix. Let $\Upsilon = R_1$.
3: Compute the Jacobian matrix at x_1, $J_1(x_1) = \frac{df(x_1)}{dx_1}$, where $x_1 = f(x_0)$.
4: Let $J_2^* = J_1 \cdot Q_1$ and decompose J_2^* using the QR decomposition: $J_2^* = Q_2 \cdot R_2$ where Q_2 is also an orthogonal matrix and R_2 is an upper triangular matrix. Let $\Upsilon = R_2 \cdot \Upsilon$.
5: Repeat Steps 3 and 4 $t-1$ times till we finally obtain $\Upsilon = R_t R_{t-1} \cdots R_1$.
6: The LE spectrum is defined by the logarithm of the diagonal elements v_{ii} of Υ: $\lambda_i = \frac{1}{t} \log |\Upsilon_{ii}|$.

As we know from the records of [2, 5], the external stimulus a is considered as a constant, i.e., $a = 2$. In [2, 5], the authors have shown the LE spectrum of Equation (16) while a varies from 0 to 1. However, for a single Adachi neuron, the role of a has not yet been discussed. As a matter of fact, most of papers which concern the Lyapunov analysis of the AdNN and it's variants focus on the parameter k_f and k_r [2, 15, 6, 18, 24]. Indeed, the role of α and a have never been discussed seriously in the literature.

Figures 6 (a) and (b) show the largest LE spectrum as obtained by Equations (31)-(33). We can clearly observe that the largest LE fluctuates sharply with a and α: For some values of a and α, the largest LE is positive and for others it is negative. From this, we can understand that in all of the papers [2, 15, 6, 18, 24], the values of $a = 2$ and $\alpha = 10$ were really set arbitrarily. Indeed, these parameters could have been replaced by one of many other values. As a matter of fact, we have verified in a experimental way that the AdNN possesses *more powerful* AM properties when $a = 7$. As we can see from Figures 6 (a) and (b), when $a = 2$ and $\alpha = 10$, the largest LE is negative which indicates that there is no chaos. This also explains why the AdNN converges after a long transient phase — which is a phenomenon also reported in [2] — approximately $21,000$ iterations[8].

4.2 Lyapunov Analysis of the Ideal-M-AdNN

We shall now perform an analysis of the Ideal-M-AdNN (i.e., the entire network) from the viewpoint of its Lyapunov Exponents.

As we stated earlier in Section 3, the Ideal-M-AdNN originates from the AdNN and the M-AdNN. It differs from the M-AdNN in its parameter settings w_{ij}, ε and a_i, which is quite crucial, because the same network can exhibit

---

[8] Elsewhere, we have verified that the size of transient phase varies with the actual initial input[25].

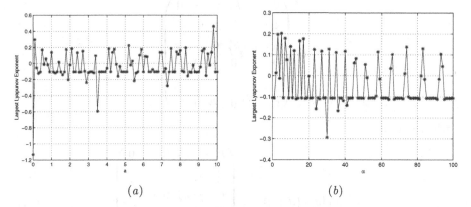

Fig. 6. The variation of the Largest Lyapunov Exponents of a single neuron with the parameters a and α. These two figures are plotted using Algorithm 1. In (a), a varies from 0 to 10 with a step size of 0.1. In (b), α varies from 0 to 100 with a step size of 1.

completely different phenomena dependent on the settings themselves. Consequently, the theoretical Lyapunov analysis of the Ideal-M-AdNN should be exactly the same as that of the M-AdNN, except that the LEs are evaluated at the *current* parameter settings.

As demonstrated in [6], the Jacobian matrix J of the Ideal-M-AdNN (and the M-AdNN) can be seen to have the form[9]:

$$
J = \begin{pmatrix} J_{ij}^1 & J_{ij}^2 \\ J_{ij}^3 & J_{ij}^4 \end{pmatrix} = \begin{pmatrix} 0 \cdots 0 \, k_f \, 0 \, \cdots 0 \\ \vdots \, \ddots \, \vdots \, \vdots \, \vdots \, \ddots \, \vdots \\ 0 \cdots 0 \, k_f \, 0 \, \cdots 0 \\ 0 \cdots 0 \, 0 \, \cdots 0 \, \, k_r \\ \vdots \, \ddots \, \vdots \, \vdots \, \ddots \, \vdots \, \vdots \\ 0 \cdots 0 \, 0 \, \cdots 0 \, \, k_r \end{pmatrix} = \begin{cases} k_f, \, 0 < i \le N, \quad j = N \\ k_r, \, N < i \le 2N, j = 2N \\ 0, \quad otherwise \end{cases}
$$

where J_{ij}^n is an N by N matrix, and $J_{ij}^1(t) = \frac{\partial \eta_i(t+1)}{\partial \eta_j(t)}$, $J_{ij}^2(t) = \frac{\partial \eta_i(t+1)}{\partial \xi_j(t)}$, $J_{ij}^3(t) = \frac{\partial \xi_i(t+1)}{\partial \eta_j(t)}$, $J_{ij}^4(t) = \frac{\partial \xi_i(t+1)}{\partial \xi_j(t)}$ respectively.

Generally speaking, it is not easy to compute the limit of Equation (28). However, due to the special form of J that we encounter here, we are able to calculate the value of Λ easily. As illustrated above,

---

[9] For the interest of brevity, we omit the detailed algebraic steps, but only present the final results. The details of these expressions can be found in [6]. Since we have chosen the Nth neuron as the global neuron, we see that only the Nth column vector of J_{ij}^1 and J_{ij}^4 is non-zero.

$$J_t = \begin{pmatrix} 0 \cdots 0 & k_f^t & 0 & \cdots & 0 \\ \vdots & \ddots & \vdots & \vdots & \vdots & \ddots & \vdots \\ 0 \cdots 0 & k_f^t & 0 & \cdots & 0 \\ 0 \cdots 0 & 0 & \cdots & 0 & k_r^t \\ \vdots & \ddots & \vdots & \vdots & \ddots & \vdots & \vdots \\ 0 \cdots 0 & 0 & \cdots & 0 & k_r^t \end{pmatrix}$$

whence we get:

$$\Lambda = \lim_{t \to \infty} [J_t \cdot (J_t)^T]^{\frac{1}{2t}} = \begin{pmatrix} k_f & \cdots & k_f & 0 & \cdots & 0 \\ \vdots & \ddots & \vdots & \vdots & \ddots & \vdots \\ k_f & \cdots & k_f & 0 & \cdots & 0 \\ 0 & \cdots & 0 & k_r & \cdots & k_r \\ \vdots & \ddots & \vdots & \vdots & \ddots & \vdots \\ 0 & \cdots & 0 & k_r & \cdots & k_r \end{pmatrix}. \qquad (34)$$

By a simple algebraic analysis we see that Λ has three different eigenvalues: Nk_f, Nk_r and 0. As a result, the Lyapunov Exponents are:

$$\lambda_1 = \cdots \lambda_{N-1} = -\infty,$$

$$\lambda_N = \log N + \log k_f > 0,$$

$$\lambda_{N+1} = \cdots \lambda_{2N-1} = -\infty,$$

$$\lambda_{2N} = \log N + \log k_r > 0.$$

In conclusion, the Ideal-M-AdNN has two positive LEs, which indicates that the network is truly a chaotic network!

At this juncture, we would like to point out that in the work of [6], the authors took another form of Equation (28) as

$$\Lambda = \lim_{t \to \infty} [(J_t)^T \cdot J_t]^{\frac{1}{2t}}. \qquad (35)$$

Consequently, the Largest LE is the maximum of the $\{\frac{1}{2} \log N + \log k_f, \frac{1}{2} \log N + \log k_r\}$, which is different from ours. However, from a realistic perspective, the choice of the definition does not matter. Indeed, independent of the definition we use, the result is consistent: The largest LE is positive which implies that the network is chaotic.

It's very interesting to compare this result with the one presented for the AdNN. Indeed, as we can see from [26], the AdNN has two different LEs: $\log k_f$ and $\log k_r$. Thus the largest LE is $\max\{\log k_f, \log k_r\}$. One can observe the following from Figure 7 (a): When $k_r < 0.2$, the largest LE is approximately -1.6, while $k_r > 0.2$, the largest LE is[10] $\log k_r$. As opposed to this, the largest LE of

---

[10] Please note that due to the fact that the computation is done numerically, there are some numerical errors, and thus the largest LE is not exactly, but very close to $\log k_r$ at some points.

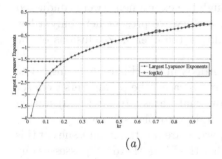 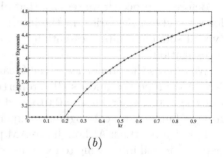

(a) (b)

Fig. 7. The spectrum of the largest LEs of (a) the AdNN and of (b) the Ideal-M-AdNN, in which we vary k_r from 0 to 1 with a step size of 0.02, and where $k_f = 0.2$. Figure (a) is calculated using Algorithm 1 while (b) is calculated using Algorithm 2.

the Ideal-M-AdNN is always positive, as shown in (b). The difference is that by binding the states of all the neurons to a single *"global"* neuron, we force the Ideal-M-AdNN to have two positive LEs.

5   Chaotic and PR Properties of the Ideal-M-AdNN

We shall now report the properties of the Ideal-M-AdNN. The protocol of our experiments is quite simple: If the network output resonates a known input pattern with *any* periodicity, we deem this pattern to have been recognized from a chaotic PR perspective. Otherwise, we say that it is unrecognizable. These properties have been gleaned as a result of examining the Hamming distance between the input pattern and the patterns that appear at the output. As a comparison, we also list the simulation results of the M-AdNN. By comparing the Hamming distance of the Ideal-M-AdNN and the M-AdNN, we can then conclude which of the schemes performs better when it concerns PR properties. In this regard, we mention that the experiments were conducted using two data sets, namely the figures used by Adachi *et al* given in Figure 8 (a), and the numeral data sets used by Calitoiu *et al* [15, 6] given in Figure 8 (b). In both the cases, the patterns were described by 10×10 pixel images, and the networks thus had 100 neurons.

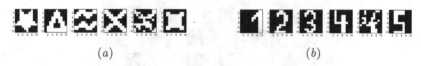

(a) (b)

Fig. 8. The 10×10 patterns used by Adachi *et al* (on the top) and Calitoiu *et al* (at the bottom). In both figures (a) and (b), the first four patterns are used to train the network. The fifth patterns are obtained from the corresponding fourth patterns by including 15% noise in (a) and (b) respectively. In each case, the sixth patterns are the untrained patterns.

Before we proceed, we remark that although the experiments were conducted for a variety of scenarios, in the interest of brevity, we present here only a few typical sets of results – essentially, to catalogue the overall conclusions of the investigation.

We discuss the properties of the Ideal-M-AdNN in two different settings: The NNs AM properties and its PR properties. In all cases, the parameters were set to be $k_f = 0.2$, $k_r = 0.9$, $\varepsilon = 0.015$, and all internal states $\eta_i(0)$ and $\xi_i(0)$ started from 0. Further, we catalogue our experimental protocols as follows:

1. AM properties: Although the AM properties are not the main issues of this paper, it is still interesting to examine whether the Ideal-M-AdNN possesses any AM-related properties. This is because the Ideal-M-AdNN is a modified version of the AdNN and the M-AdNN. By comparing the differences between the Ideal-M-AdNN and the AdNN or the M-AdNN, we can obtain a better understanding of its dynamical properties. In this case, the external stimulus a is a constant, i.e., $a = 2$.

2. PR properties: This investigation is really the primary intent of this paper. In this case, we investigate whether our new network is able to achieve pattern recognition. This is accomplished by checking whether the output can respond correctly to different inputs. In our experiments, we tested the network with known patterns, noisy patterns and with unknown patterns. In this case, the external stimulus was set to $a = 2 + 6P$, where P is the input pattern.

5.1   AM Properties

We now examine whether the Ideal-M-AdNN possesses any AM-related properties for certain scenarios, i.e., if we fix the external input $a_i = 2$ for all neurons. The observation that we report is that during the first 1,000 iterations (due to the limitations of the file size, we present here only the first 36 images), the network only repeats black and white image. This can be seen in Figure 9.

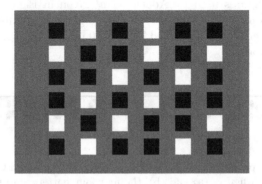

Fig. 9. The visualization of the output of the Ideal-M-AdNN under the external input $a_i = 2$. We see the output switches between images which are entirely only black or white.

It is very easy to understand this phenomenon: First of all, we see that all the neurons have the same output, 0, at time t (i.e, $t = 0$). This is true because we start the iteration with all internal states $\eta_i(0) = 0, \xi_i(0) = 0$ and $x_i(0) = 0$. As a result, the first image of Figure 9 is completely black[11]. At the next time step, $\eta_i(1) = 0$, $\xi_i(1) = a > 0$, which causes the output of all the neurons to be $x_i(1) = f(\eta+\xi) \approx 1$, which is why we see a completely white image. At the third time step, we first computed the summation $\Sigma_{j=1}^{j=N} w_{ij} x_j(1)$ where w_{ij} is defined by Equation (15). We can verify that the summation of each line of of the matrix w is either -0.5 or 0.5. Thus, $\eta_i(2)$ is either 0.5 or -0.5. Meanwhile, we must note that $\alpha = 10$, which results in a negative value for $\xi_i(2)$ and $\eta_i(2) + \xi_i(2)$, and consequently, $x_i(1) = f(\eta+\xi) \approx 0$. This is the reason why we see the third image of Figure 9 to be completely black. If we follow the same arguments, we see that at any time instant, the outputs of all the neurons only switch between 1 and 0 synchronically, which means the output image of Figure 9 switches between black and white, implying that the network possesses *no* AM properties at all.

As a comparison, we mention that the M-AdNN also does not possess any AM-related properties. In this regard, the M-AdNN and the Ideal-M-AdNN are similar.

5.2   PR Properties

The PR properties of the Ideal-M-AdNN are the main concern of this paper. As illustrated in Section 3, the goal of a chaotic PR system is the following: The system should respond periodically to trained input patterns, while it should respond chaotically (with chaotic outputs) to untrained input patterns. We now confirm that the Ideal-M-AdNN does, indeed, possess such phenomena.

We now present an in-depth report of the PR properties of the Ideal-M-AdNN's by using a Hamming distance-based analysis. The parameters that we used were: $k_f = 0.2$, $k_r = 0.9$, $\varepsilon = 0.015$ and $a_i = 2+6x_i$. The PR-related results of the Ideal-M-AdNN are reported for the three scenarios, i.e., for trained inputs, for noisy inputs and for untrained (unknown) inputs respectively. We report only the results for the setting when the original pattern and the noisy version are related to P4. The results obtained for the other patterns are identical, and omitted here for brevity.

1. The external input of the network corresponds to a known pattern, say P4. To report the results for this scenario, we request the reader to observe Figure 10, where we can find that P4 is retrieved periodically as a response to the input pattern. This occurs 391 times in the first 500 iterations. On the other hand, the other three patterns never appear in the output sequence. The phase diagrams of the internal states that correspond to Figure 10 are shown in Figure 11, where the x-axes are $\eta_{86}(t) + \xi_{86}(t)$ (the neuron with the index 86 has been set to be the global neuron), and where the y-axes are $\eta_i(t) + \xi_i(t)$ where the index $i = 80, \cdots, 88$ (i is chosen randomly) respectively.

---

[11] In our visualization, the value 0 means "black" while 1 means "white".

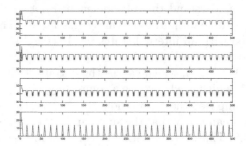

Fig. 10. PR properties: The Hamming distance between the output and the trained patterns. The input was the pattern P4. Note that P4 appears periodically, i.e., the Hamming distance is zero at periodic time instances.

From Figures 10 and 11, we can conclude that the input pattern can be recognized periodically if it is one of the known patterns. Furthermore, from Figure 11, we verify that the periodicity is 14, because all the phase plots have exactly 14 points.

As a comparison, we report that the M-AdNN is also able to recognize known patterns under *certain* conditions. However, this is not a phenomenon that we can "universally" ascribe to the M-AdNN because it fails to pass the necessary tests!

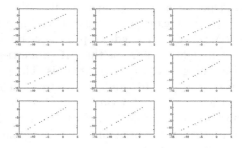

Fig. 11. PR properties: The phase diagrams of the internal states corresponding to Figure 10

2. The external input of the network corresponds to a noisy pattern, in this case P5, which is a noisy version of P4.

   Even when the external stimulus is a garbled version of a known pattern (in this case P5 which contains 15% noise), it is interesting to see that *only* the original pattern P4 is recalled periodically. In contrast, the others three known patterns are *never* recalled. This phenomenon can be seen from the Figure 12. By comparing Figures 10 and 12, we can draw the conclusion that the Ideal-M-AdNN can achieve chaotic PR even in the presence of noise and distortion.

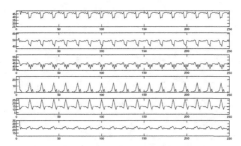

Fig. 12. PR properties: The Hamming distance between the output and the trained patterns. The input was the pattern P5. Note that P4 (not P5) appears periodically, i.e., the Hamming distance is zero at periodic time instances.

As in the previous case, the phase diagrams of the internal states that correspond to Figure 12 are shown in Figure 13, where the x-axes are $\eta_{86}(t)+\xi_{86}(t)$ (the neuron with the index 86 has been set to be the global neuron), and where the y-axes are $\eta_i(t) + \xi_i(t)$ where the index $i = 80, \cdots, 88$ (i is chosen randomly) respectively. As before, from Figure 13, we verify that the periodicity is 14, because all the phase plots have exactly 14 points.

Indeed, even if the external stimulus contains some noise, the Ideal-M-AdNN is still able to recognize it correctly, by resonating periodically! Furthermore, we have also tested the network using noisy patterns which contained up to 33% noise. We are pleased to report that it still resonates the input correctly, as shown in Figure 14. If the noise is even higher, i.e., more than 34%, the PR properties tend to gradually disappear. In this case, understandably, the simulation results are almost the same as when does the testing with *unknown* patterns.

3. The external input of the network corresponds to an unknown pattern, P6. In this case we investigate whether the Ideal-M-AdNN is capable of distinguishing between known and unknown patterns. Thus, we attempt to

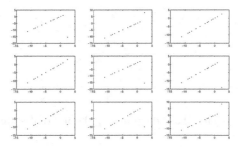

Fig. 13. PR properties: The phase diagrams of the internal states corresponding to Figure 12

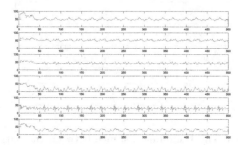

Fig. 14. The PR properties of the Ideal-M-AdNN when it encounters a noisy pattern which contain up to 33% noise. Observe that the system still resonates the corresponding known pattern correctly.

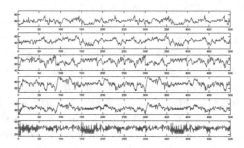

Fig. 15. PR properties: The Hamming distance between the output and the trained patterns. The input pattern was P6.

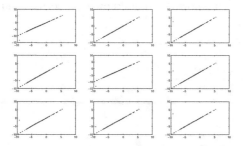

Fig. 16. PR properties: The phase diagrams of the internal states corresponding to Figure 15

stimulate the network with a completely unknown pattern. In our experiments, we used the pattern P6 of Figure 8 (a) initially used by Adachi *et al.* From Figure 15 we see that neither those known patterns nor unknown pattern appear.

As in the previous two cases, the phase diagrams of the internal states that correspond to Figure 15 are shown in Figure 16, where the x-axes are $\eta_{86}(t) + \xi_{86}(t)$ (where, as before, the neuron with the index 86 has been set to be the global neuron), and where the y-axes are $\eta_i(t) + \xi_i(t)$ where the index $i = 80, \cdots, 88$ (i is chosen randomly) respectively. The lack of periodicity can be observed from Figure 16, since the plots themselves are dense.

By way of comparison, we can see that the M-AdNN fails to pass this test, as can be seen from Figure 17. In this figure, the input is an unknown pattern, P6. We expect the network to respond chaotically to this input. Unfortunately, one sees that it still yields a periodic output. In short, no matter what the input is, the M-AdNN always yields the input pattern itself, and that, *periodically*. Obviously, this property cannot be deemed as PR. As opposed to this, the Ideal-M-AdNN responds intelligently to the various inputs with correspondingly different outputs, each resonating with the input that excites it – which is the crucial "golden" hallmark characteristic of a *Chaotic* PR system. Indeed, the switch between "order" (resonance) and "disorder" (chaos) seems to be consistent with Freeman's biological results – which, we believe, is quite fascinating!

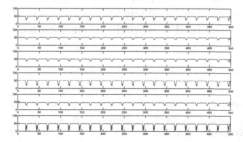

Fig. 17. The Hamming distance between the output and the trained patterns of the M-AdNN. The output repeats the input pattern, P6, periodically.

6   Conclusions

In this paper we have concentrated on the field of Chaotic Pattern Recognition (PR), which is a relatively new sub-field of PR. Such systems, which have only recently been investigated, demonstrate chaotic behavior under normal conditions, and "resonate" (i.e., by presenting at the output a specific pattern frequently) when it is presented with a pattern that it is trained with. This ambitious goal, with the requirement of switching from chaos to periodicity is, indeed, most demanding.

While the Adachi Neural Network (AdNN) [1–5] has properties which are pseudo-chaotic, it also possesses *limited* PR characteristics. As opposed to this, the Modified Adachi Neural Network (M-AdNN) proposed by Calitoiu *et al* [6], is a fascinating NN which has been shown to possess the required periodicity property desirable for PR. In this paper, we have explained why the PR properties claimed in [6] are not as powerful as originally claimed. Thereafter, we have argued for the basis for setting the parameters of the M-AdNN. By appropriately tuning the parameters of the M-AdNN for its weights, steepness and external inputs, we have obtained a new NN which we have called the Ideal-M-AdNN. Using a rigorous Lyapunov analysis, we have demonstrated the chaotic properties of the Ideal-M-AdNN. We have also verified that the system is truly chaotic for untrained patterns. But most importantly, we have shown that it is able to *switch to being periodic* whenever it encounters patterns with which it was trained (or noisy versions of the latter).

Apart from being quite fascinating, as far as we know, the theoretical and experimental results presented here are both unreported and novel. However, as the reader can observe, we were not so concerned about the application domain. Rather, our work was to be of an investigatory sort. Thus, although the data sets that we used were rather simplistic, they were used to submit a *prima facie* case. On one hand, if we could achieve our goal with these data sets, we believe that this research direction has the potential of initiating new research avenues in PR and AM. On the other hand, from a practical perspective, we believe that this is also beneficial for other computer science applications such as image searching, content addressed memory etc., and this is considered open at present. However, although we have not yet developed any immediate applications, we are willing to collaborate with others (and also provide them with our code) to do this.

References

1. Adachi, M., Aihara, K., Kotani, M.: An analysis of associative dynamics in a chaotic neural network with external stimulation. In: International Joint Conference on Neural Networks, Nagoya, Japan, vol. 1, pp. 409–412 (1993)
2. Adachi, M., Aihara, K.: Associative dynamics in a chaotic neural network. Neural Networks 10(1), 83–98 (1997)
3. Adachi, M., Aihara, K.: Characteristics of associative chaotic neural networks with weighted pattern storage-a pattern is stored stronger than others. In: The 6th International Conference On Neural Information, Perth Australia, vol. 3, pp. 1028–1032 (1999)
4. Adachi, M., Aihara, K., Kotani, M.: Pattern dynamics of chaotic neural networks with nearest-neighbor couplings. In: The IEEE International Sympoisum on Circuits and Systems, vol. 2, pp. 1180–1183. Westin Stanford and Westin Plaza, Singapore (1991)
5. Aihara, K., Takabe, T., Toyoda, M.: Chaotic neural networks. Physics Letters A 144(6-7), 333–340 (1990)
6. Calitoiu, D., Oommen, B.J., Nussbaum, D.: Periodicity and stability issues of a chaotic pattern recognition neural network. Pattern Analysis and Applications 10(3), 175–188 (2007)

7. Freeman, W.J.: Tutorial on neurobiology: from single neurons to brain chaos. International Journal of Bifurcation and Chaos in Applied Sciences and Engineering 2, 451–482 (1992)
8. Pyragas, K.: Continuous control of chaos by self-controlling feedback. Physics Letters A 170(6), 421–428 (1992)
9. Nozawa, H.: A Neural-Network Model as a Globally Coupled Map and Applications Based on Chaos. Chaos 2(3), 377–386 (1992)
10. Chen, L., Aihara, K.: Chaotic simulated annealing by a neural network model with transient chaos. Neural Networks 8(6), 915–930 (1995)
11. Tsui, A.P.M., Jones, A.J.: Periodic response to external stimulation of a chaotic neural network with delayed feedback. International Journal of Bifurcation and Chaos 9(4), 713–722 (1999)
12. Hiura, E., Tanaka, T.: A chaotic neural network with duffing's equation. In: Proceedings of International Joint Conference on Neural Networks, Orlando, Florida, USA, pp. 997–1001 (2007)
13. Tanaka, T., Hiura, E.: Computational abilities of a chaotic neural network. Physics Letters A 315(3-4), 225–230 (2003)
14. Tanaka, T., Hiura, E.: Dynamical behavior of a chaotic neural network and its applications to optimization problem. In: The International Joint Conference On Neural Network, Montreal, Canada, pp. 753–757 (2005)
15. Calitoiu, D., Oommen, B.J., Nussbaum, D.: Desynchronizing a chaotic pattern recognition neural network to model inaccurate perception. IEEE Transactions on Systems Man and Cybernetics Part B-Cybernetics 37(3), 692–704 (2007)
16. Qin, K., Oommen, B.J.: Chaotic pattern recognition: The spectrum of properties of the adachi neural network. In: da Vitoria Lobo, N., Kasparis, T., Roli, F., Kwok, J.T., Georgiopoulos, M., Anagnostopoulos, G.C., Loog, M. (eds.) S+SSPR 2008. LNCS, vol. 5342, pp. 540–550. Springer, Heidelberg (2008)
17. Qin, K., Oommen, B.J.: An enhanced tree-shaped adachi-like chaotic neural network requiring linear-time computations. In: The 2nd International Conference on Chaotic Modeling, Simulation and Applications, Chania, Greece, pp. 284–293 (2009)
18. Qin, K., Oommen, B.J.: Adachi-like chaotic neural networks requiring linear-time computations by enforcing a tree-shaped topology. IEEE Transactions on Neural Networks 20(11), 1797–1809 (2009)
19. Nagumo, J., Sato, S.: On a response characteristic of a mathematical neuron model. Biological Cybernetics 10(3), 155–164 (1971)
20. Eckmann, J.P., Rulle, D.: Ergodic theory of chaos and strange attractors. Reviews of Modern Physics 57(3), 617–656 (1985)
21. Sprott, J.C.: Numerical calculation of largest lyapunov exponent (1997)
22. Wolf, A., Swift, B.J., Swinney, L.H., Vastano, A.J.: Determining lyapunov exponent from a time series. Physica 16D, 285–317 (1985)
23. Sandri, M.: Numerical calculation of lyapunov exponents. The Mathematica Journal, 78–84 (1996)
24. Qin, K., Oommen, B.J.: Networking logistic neurons can yield chaotic and pattern recognition properties. In: IEEE International Conference on Computational Intelligence for Measure Systems and Applications, Ottawa, Ontario, Canada, pp. 134–139 (2011)
25. Luo, G.C., Ren, J.S., Qin, K.: Dynamical associative memory: The properties of the new weighted chaotic adachi neural network. IEICE Transactions on Information and Systems E95d(8), 2158–2162 (2012)
26. Adachi, M., Aihara, K.: An analysis on instantaneous stability of an associative chaotic neural network. International Journal of Bifurcation and Chaos 9(11), 2157–2163 (1999)

A Framework for an Adaptive Grid Scheduling: An Organizational Perspective

Inès Thabet[1,2], Chihab Hanachi[1], and Khaled Ghédira[2]

[1] Institut de Recherche en Informatique de Toulouse IRIT, UMR 5505,
Université Toulouse 1 Capitole, 2 rue du Doyen-Gabriel-Marty, 31042 Toulouse Cedex 9
{Ines.Thabet,Chihab.Hanachi}@irit.fr
[2] Stratégie d'Optimisation et Informatique IntelligentE,
ISG Tunis, 41 Rue de la Liberté, Cité Bouchoucha 2000 Le Bardo, Tunis-Tunisie
Khaled.Ghedira@isg.rnu.tn

Abstract. Grid systems are complex computational organizations made of several interacting components evolving in an unpredictable and dynamic environment. In such context, scheduling is a key component and should be adaptive to face the numerous disturbances of the grid while guaranteeing its robustness and efficiency. In this context, much work remains at low-level focusing on the scheduling component taken individually. However, thinking the scheduling adaptiveness at a macro level with an organizational view, through its interactions with the other components, is also important. Following this view, in this paper we model a grid system as an agent-based organization and scheduling as a cooperative activity. Indeed, agent technology provides high level organizational concepts (groups, roles, commitments, interaction protocols) to structure, coordinate and ease the adaptation of distributed systems efficiently. More precisely, we make the following contributions. We provide a grid conceptual model that identifies the concepts and entities involved in the cooperative scheduling activity. This model is then used to define a typology of adaptation including perturbing events and actions to undertake in order to adapt. Then, we provide an organizational model, based on the Agent Group Role (AGR) meta-model of Freber, to support an adaptive scheduling at the organizational level. Finally, a simulator and an experimental evaluation have been realized to demonstrate the feasibility of our approach.

Keywords: Grid Scheduling, Multi-Agent System, Adaptation.

1 Introduction

1.1 Context

Grid computing [1] provides computing capacities to high performance and data intensive applications by taking advantage of the power of widely distributed and heterogeneous resources.

Grids may be viewed as complex computational organizations. Indeed, grids are made of a large number of *heterogeneous, distributed* and *autonomous* components

N.T. Nguyen (Ed.): Transactions on CCI XI, LNCS 8065, pp. 52–87, 2013.

(administrative domains, network links, grid schedulers, local scheduler, resources, etc.) that have to interact frequently and cooperate to perform efficiently user applications. Without regulation, these interactions can become numerous, costly and unpredictable and therefore lead to a chaotic collective behaviour.

Regulation and efficiency are mainly dependent on the *grid scheduling* (GS) system and its interactions with the other components. The grid scheduling system allocates application's tasks to the most efficient resources at the appropriate time. Its interactions are also of paramount importance for finding available resources and allocating them tasks.

Adaptiveness is a required quality of the grid scheduling system since it has to face numerous disturbances due to the grid dynamicity. This dynamicity is due to the open character of the grid (resources can join or leave at any moment) and to performances variability of the resources: the network links or compute nodes may become overloaded or unavailable because of crashes or because they have been claimed by a prior (higher priority) application. Also, a more suitable resource may become available through time, etc.

1.2   The Problem Being Addressed

Giving this context, as detailed in the related work section, much work has been developed to design and implement adaptive grid scheduling. However, they remain at low-level, focusing on the scheduling component taken individually and modifying its structure and behaviour. This kind of ad-hoc solutions is not adequate since it is difficult to control and act on the internal behaviour of each component (micro-level) which are dynamic, numerous and, above all, unknown a priori.

Here, we argue that thinking the scheduling adaptiveness at a macro level, through the components organizations and interactions, is important and realistic and could better capture the grid complexity. Following this view, this paper adopts an agent-based organization perspective to design and simulate grid scheduling as a cooperative activity. Indeed, agent technology provides high level organizational concepts (groups, roles, commitments, interaction protocols) to structure, coordinate and ease the adaptation of distributed systems efficiently.

Giving these observations, the question addressed is: "how to design and simulate *an adaptive grid scheduling system* at *a macro level* (regulation of the components) according to an agent organizational perspective?"

1.3   An Agent-Based Organizational Approach

This paper deals with the modelling and simulation of *an adaptive grid scheduling* system able to detect the environment disturbances and to respond to dynamic changes in an efficient way. In fact, grid scheduling adaptation can be considered according to different views and requires to identify *i) the perturbing events*, *ii) the objects/components* that could be subject to adaptation, and *iii) the actions* to undertake in order to adapt.

From the conventional grid scheduling system [2][3], we have identified the *four following perspectives* to be considered in order to perform the adaptation:

- *The environment.* The grid environment is mainly composed of grid resources and applications. The environment adaptation is considered according to two aspects. The first aspect concerns the grid resources that could be aware of perturbations events and be able to perform adaptation actions such as rejecting a submitted task when it is overloaded, preempting a prior task, sending a failure notification, etc. The second aspect is about the submitted applications that could adapt to perturbations when the scheduling components is integrated in the application itself [4][5][6]. In this case, each application has its own scheduler that determines a performance-efficient schedule, monitors resources performances at the execution time and undertakes adequate adaptation action in case of disturbances.
- *The software components.* Grid scheduling components have to be endowed with the capacity to monitor the execution of tasks and to adapt in the case of disturbances such as rescheduling a failed task, scheduling policy adaptation, etc.
- *The grid organization.* Grid resources are physically organized in clusters, sites, domains, etc. One or more grid resources can also form a virtual organization [7] to offer new or better quality of service. In this context, the grid could be considered as a flexible computational organization made of interacting components: any change in the environment can be easily translated as reorganization, role modification, use of different interaction protocols, etc.
- *The interactions between grid components.* Interactions are here crucial for adaptation since they allow the grid components to cooperate efficiently to find an adequate partner to execute optimally application's tasks even if a perturbation occurs (use of contract net, auction protocols, etc.).

Giving these views, we define a typology of the adaptation in the grid (see section 3) covering the four previous views. Then, we focus on the organizational view following an agent-based approach.

The agent-based organizational approach is a conceptual framework providing the advantages of abstracting a system at a high level with macro and social concepts (roles, interaction protocols, groups) and allows the easy design and implementation of open and dynamic systems such as grids. Moreover, organizations are flexible entities able to adapt their design to any change in their environment and techniques to make them adaptive are today available [8] and could be used with benefit for the adaptation of the grid scheduling. *Multi-agent systems (MAS),* widely recognized in the grid literature [9], also provide abstractions to build autonomous, reactive and proactive software components able to communicate with sophisticated languages and protocols. Not only these concepts should help us to better structure and regulate the components behaviour, their interactions and their execution but also it should improve the efficiency of the grid scheduling by structuring the high number of communications and by supporting the grid organization.

1.4   Our Contributions

The contribution of this paper is the definition of a *framework for designing and simulating a multi-agent organizational model for an adaptive grid scheduling*. Our contributions consist of four components:

- *A conceptual model of the whole grid system (section 2) and its related typology of adaptation (section 3).* Our grid modelling includes both a conceptual model and a multi-agent model of the grid. The first model represents an abstraction of the grid system in which the concepts, their properties, and their relations are identified. This representation is computation independent and is used to identify i) a number of situations requiring an adaptation (according to the objects of the conceptual models) such as resource breakdown, departure, overload, etc. ii) a number of actions to undertake to adapt to disturbances (application migration, rescheduling, change of the scheduling policy, etc.). The second model describes the representation of the first one in terms of agents. In fact, our grid components are fully rethought and modelled as agents in order to implement an *effective distributed and coordinated adaptive scheduling*. Our multi-agents model is represented according to the "vowel" approach [10] that describes the MAS as a set of four main perspectives: the environment, the software components (agents) involved in the system, their interactions and their organization. The proposed multi-agents model is platform independent since it is not tied to any particular implementation and is used to describe a general multi-agents architecture for an adaptive grid scheduling.
- *An adaptation model for grid scheduling.* This model follows an agent-based organizational perspective. It describes the actors involved in the system, the links that exist between them and the protocols that rule their interactions. In this model, the adaptation is supported by reorganization and by the flexibility of the interaction protocols. This model follows the Agent Group Role (AGR) meta-model of Ferber [11] and is platform dependent.
- *A simulator of the adaptation model.* This simulator is compliant with our organizational model and it simulates any organization represented by our model. The implementation is based on the Madkit platform [12] that integrates the AGR meta-model.
- *Performance criteria definition and an evaluation of our adaptation model (section 6).* The performance criteria are related to the organization qualities. They are adapted from Grossi [13] and Kaddoum [14] and include structural and statistical aspects. The *structural aspect* is concerned with the network topology and is based on graph theory's measures. The *statistical aspect* is concerned with the communication load and with the execution time.

1.5   Organization of the Paper

The remainder of this paper is organized as follows. In section 2, we describe our grid conceptual model and our multi-agent grid model according to the environment, agent, organization and interaction views. We also specify accurately in section 2 the components behaviour and the protocols ruling their cooperation using Petri Nets [15]. In section 3, we introduce a typology of disturbing events and the type of adaptation that could be undertaken. An overview of the proposed organizational model for grid scheduling (AGR model, protocols, groups, roles, etc.) is presented in section 4. In section 5, we describe the developed simulator used to evaluate our model experimentally. We propose performance criteria and the evaluation of our organizational model for grid scheduling in section 6. In section 7, we review the related work. Finally, we summarize and lay out the future work.

2   Conceptual Modelling of the Grid Scheduling

This section presents the definition of the concepts and the different entities involved in the grid scheduling system. To this end two models are introduced: a *conceptual model* and a *multi-agent model* of our grid scheduling system. The *first* one represents a high level abstraction of our domain that is independent of any implementation technology. This model has been designed in order to give a description of our universe of discourse and is mainly used in order to define an adaptation typology (see section 3). The *second* one represents the agent approach adopted by our work. To this end, the conceptual model of the grid scheduling is modelled as a multi-agent system that specifies how the functionalities of our grid are realized using agents. This model is platform independent. Finally, a multi-agent organizational model that represents our developed framework is introduced in section 4.

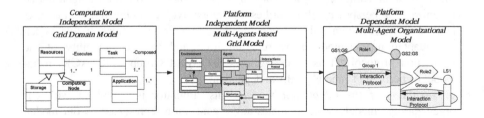

Fig. 1. Conceptual modelling of the grid scheduling

2.1   A Conceptual Model of the Grid Scheduling

In this section, we describe our domain which is a representation of the concepts implied in the grid scheduling and their relations. They are represented using UML (Unified Modelling Language) notation.

The concepts identified and the properties mentioned were selected following a deep investigation of the domain [2][3][16]. The instances of these models define the objects that can be subject to adaptation and the actions to undertake in order to adapt in case of disturbances.

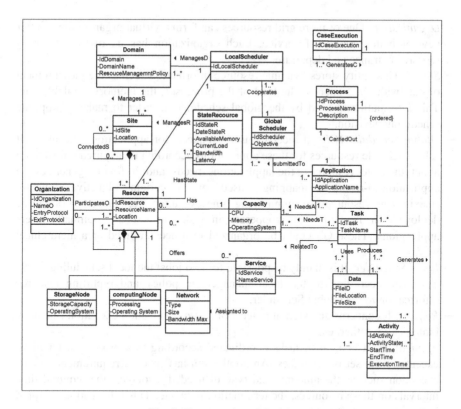

Fig. 2. Conceptual model of the grid

The final model is depicted in figure 2. The most important concepts of this model are:

— *Site.* It is an autonomous entity composed of one or several resources and managed by an administrative domain in which the local resource management policy is well specified.

— *Resource.* It is the central class of our model. According to [16], a resource is a basic device where jobs are scheduled/processed/assigned. Resources are required to execute users' applications. The resource can be a processor responsible for tasks processing, a data storage in which the data needed for tasks execution is stored, or a network link used for data transfer. Each resource can be connected to one or more other resources and has both static and dynamic characteristics. Static characteristics are about the resource name, the memory size, the operating system, etc. Dynamic characteristics concern information that can evolve over time such as the as the resource processing speed or load capacity.

— *Service.* Resources can be endowed with some capabilities (skills), called "Service" that are needed to execute a task. The quality of the offered service is dependent on the resource capacity (memory, CPU speed, etc.).

— *Organization*. One or more grid resources can form a virtual organization to offer new or better quality of service. Each organization has its own protocol of resource entrance and departure.

— *StateR*. This entity stores dynamic resource information such as the measurement of bandwidth and latency, the load of the processor, the memory available, etc. This information is used by the global scheduler in order to make appropriate scheduling decision.

— *Global Scheduler (GS)*. It receives users' requests for applications execution. It selects feasible resources for these applications according to acquired information about the grid resources, the applications needs and it finally generates an application-to-resource mapping based on a certain objective function (maximizing resource use, minimizing time execution, etc). Several GS can be deployed on the grid and cooperate in order to execute optimally users applications. These GS could be organized in a decentralized or a hierarchical way.

— *Local Scheduler*. It is mainly responsible for two jobs: the local scheduling inside a domain according to the local management policy and reporting resource information to the Global Scheduler.

— *Request*. It concerns an external request of a user application execution (medical, biological, weather, etc).

— *Application*. It is a set of tasks coordinated according to a process that will be executed on a set of resources. An application has specific requirements for its execution such as the amounts and type of needed resources, the required time intervals on these resources, the termination time, etc. This information is stored in the entity capacity.

— *Capacity*. It can define task or application requirements as well as the capacity of a resource. The Global Scheduler has to generate an application-to-resource mapping, based on certain objective function, resource information (its capacity and its state when the application will begin its execution on), applications and tasks information.

— *Process*. It describes the order of the application's tasks execution. From a process, several alternatives execution plans can be generated.

— *Task*. It is an atomic unit to execute on a resource. Each task has specific resources requirements for its execution (CPU, memory, beginning/finish time, etc.) stored in the entity capacity.

— *CaseExecution*. The execution of an application according to a process generates a process instance that we call "CaseExcecution".

— *Activity*. The execution of a task on a grid resource generates a task instance. This entity stores all information necessary for the adaptation, such as the task instance state (affected, waiting, in execution, stopped or finished) and the results of the execution on a given resource. These results are used by the Global Scheduler to predict the performance of task execution on a given resource.

In addition to the defined concepts, our model has an integrity constraint that we specify textually. It relates the application time execution and the sum of its tasks durations. These durations are defined through applications and tasks requirements and are stored in the class capacity. The constraint is as follows: "the application time execution must be higher or equal to the sum of its tasks execution times". This is due to the time necessary for tasks synchronization and particularly for data transfer.

2.2 A Multi-Agent Model of the Grid Scheduling

In this section we describe our multi-agent model of the grid scheduling. In fact, as we previously mentioned, all grid components are modelled as agents. Moreover, we have followed the separation of concerns principles and modelled our multi-agent system as several interacting models: the *environment*, the *agent*, the *organization* and the *interactions models*. Each model is coping with a well-defined function and is used to describe the structure and the functioning of our multi-agent system. This representation is compliant with the "Vowel" approach [10] of Demazeau that eases the construction of the multi-agent system and facilitates the reusability and the maintainability of each model. According to Demazeau:

The ***environment view*** describes the environment in which evolves the agent which may be the physical world, a user via a graphical user interface, a collection of other agents, a combination of these objects. In our case, our environment defines the physical resources with their connections, their states and capacities, the users' applications with their tasks and the agents managing the grid.

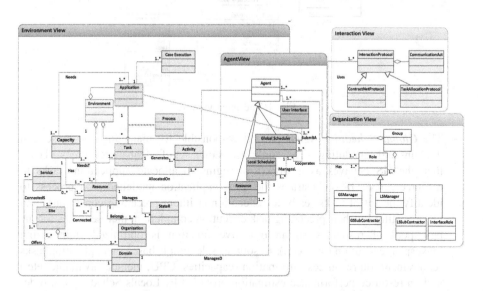

Fig. 3. Multi-agents model of the grid scheduling

The ***Agent*** **view** describes the agents, their internal architecture and their functioning. In our case, all grid components are modelled as agents. Let us introduce the agents' architecture and their functioning.

— The ***user interface (UI)*** is able to interact both with the user and with the middle layer. It submits the user query to the most adapted Global Scheduler. A user query corresponds to an application made of several coordinated tasks.

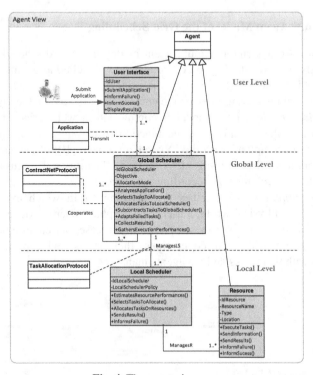

Fig. 4. The agent view

— The ***Global Scheduler Agent (GS)*** is responsible for managing one or several domains and scheduling users' applications on a set of adequate resources. To do that, the GS behaves as a workflow engine to allocate applications tasks to resources. The GS orchestrates the user' application modelled as a process to identify the set of tasks ready for execution (without preceding constraints) and tasks waiting for their preceding task execution completion, then searches for the best pool of resources so an objective function is optimized (minimizing makespan, execution time, etc.). Resources selection is based on application tasks requirement, on resources information (capacities, CPU, memory available, etc.) and on resource performance estimation provided by Locals Schedulers in order to choose the best pool of resources for application tasks execution. Tasks could therefore be allocated locally by the GS using a tasks allocation protocols or sub-contracted to GS' acquaintances (Other Global Schedulers Agents) using the Contract Net Protocol [17] if there are no adequate resources in the GS' local domains. Selected resources form a virtual organization made by multiple distributed resources for application tasks execution. The virtual organization functioning is detailed in section 4. Once tasks are submitted, the GS is also responsible for monitoring, if necessary adapting the grid functioning in case of disturbances, collecting the execution results and reporting them to the user interface. The resource monitoring concerns gathered information related to the resource executing the submitted tasks and includes resource availabilities, resource load and resource estimated completion time, etc. The adaptation mechanism is detailed in the next section and are basically based on the use of organizational structure adaptation, on rescheduling failed tasks and on interaction protocols ruling the cooperation between the grid components and allowing the applications execution completion even when perturbations occur.

— The *Local Scheduler Agent* manages one or several resources belonging to the same domain. Local Scheduler Agents receive tasks from their Global Scheduler and they are responsible for their scheduling according to a local policy. It consists in determining the order in which tasks are executed in their local domain (FIFO algorithm, greedy algorithm, etc.). The Local Scheduler Agent is also in charge of detecting failure and for reporting information to its Global Scheduler Agent to undertake adequate adaptation actions.

— Resources provide high computing capabilities to enable task execution. Each resource is managed by a *Resource Agent* in charge of reporting information (resource state and availabilities, tasks execution progress, execution results) to the Local Scheduler.

The *organization view* provides a representation of the logic agents' organization in groups in order to execute the user applications. In our case, one or multiple specialized resources (dedicated to a type of task) can form a virtual organization to offer new or better quality of service. The organizations view is modelled using the AGR (Agent Group Role) Meta-model of Ferber [11]. This model organizes the agents architecture described above using different roles and groups and will be detailed in section 4.

The *Interaction View* describes the relationship between agents through protocols or interaction language. In our architecture, the Global Scheduler Agent performs two types of protocols: a *task allocation protocol* with its Local Schedulers and the *Contract Net protocol* with other GSs. Each protocol transmits, receives and interprets communication acts. We use here FIPA-ACL [18] that supports high level communication between the different grid components. FIPA-ACL has several advantages. Each communication act follows a <performative (<message>)> form. Figure 5 shows an example of FIPA-ACL message for task execution call for proposal. Performatives transmit the intention of the communication acts (inform, query, request, call for proposal, etc.), while the message is a complex data structure where the domain and the content of the messages may refer respectively to a variable ontology and language. This feature eases interoperability between grid components and improves communication between them. Let us detail each protocol inspired from multi-agent protocols and adapted to our scheduling context. We use Agent Unified Modelling Language (AUML) to describe at a high level the interaction protocols.

```
(cfp
:sender(agent-identifier: name GS1
:receiver(set(agent-identifier: all GS
:language XML
:ontology gridModelOntology (compliant with the conceptual model of the grid see section 2.1)
:content
(<?mxl version="1.0"
<taskExecutionRequest>
<type>matrix-multiplication</type>
<size>15000</size>
<numberProcessorsMin>5</numberProcessorsMin>
<Cpufrequency>2Ghz</CpuFrequency>
<MemoryMin>512</MemoryMin>
<StartTime>t+10</StartTime>
<EndTime>t+50</EndTime>
</taskExecutionRequest>
)
```

Fig. 5. Example of FIPA-ACL message for the task execution call for proposal

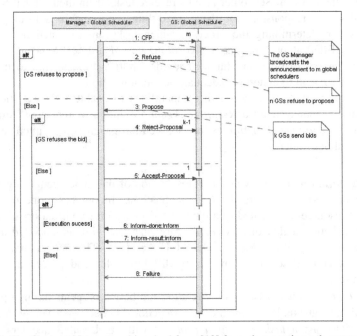

Fig. 6. The Contract Net protocol inspired from [19] for task execution sub-contracting

The *Contract Net protocol* allows an initiator Global Scheduler to sub-contract tasks to other GSs. The GS requests proposals by advertising a call for proposal using the performative *<cfp>* for task execution to its acquaintances. The initiator GS then acts as the manager of the task. The call for proposal specifies the task to be executed and the condition the manager places upon the execution. Conditions include requirements such as time or performance constraints, needed service, etc. Performance constraints describe metrics such as CPU load needed, storage capacity, bandwidth needed, etc. Time constraint describes a desirable period to complete the task execution. The constraint values may be fixed or bounded such as lower and upper bound. See example of FIPA-ACL call for proposal (Figure 5). The Global Schedulers receiving the call for proposal can either propose to execute the task under certain conditions (the price, the time, etc), or refuse to propose. The Manager receives proposals from Global Schedulers, evaluates them and chooses the Global Scheduler that maximizes its objective (to minimize time, to minimize cost, etc). An acceptance message will be sent (*<accept-proposal>*) to the chosen GS and a rejection message (*<reject-proposal>*) will be sent to the other GSs. Once the Manager sends an acceptance message, the GS has to perform task and to send back a completion message (*<inform-done*) when it performs task successfully or a message (*<failure>*) when it fails to execute the submitted task. Figure 6 shows an instantiation of the Contract net protocol [19] for task execution subcontracting. The use of the Contract Net is justified since it is one of the most flexible and efficient negotiation mechanisms and it eases tasks distribution.

The Task Allocation Protocol allows the Global Scheduler to allocate tasks to its Local Schedulers. The task allocation protocol depends on the Global Scheduler objective (minimizing the time execution, load balancing between domains, etc.). It selects the best Local Scheduler according to that objective. The LS has the possibility to accept or reject the transmitted task according to its policy, and to the state of the resources it manages. Once it has accepted a task, if some problem occurs, the LS may have to perform some adaptation. This protocol is shown in Figure 7.

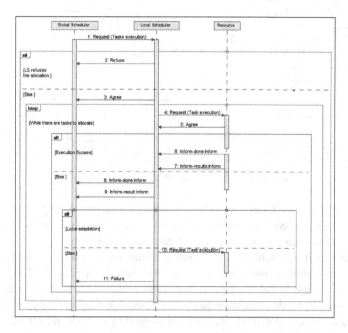

Fig. 7. The task allocation protocol (AUML sequence diagram)

2.3   Specification of Agents Behaviour and Their Interaction Protocols

The Petri Net of figure 8 provides a detailed view of interaction protocols between the Grid schedulers and their local behaviours in order to perform an application. This Petri Net shows how a Global Scheduler (called Manager) interacts with other Local and Global Schedulers. This Petri Net has been specified and simulated with the renew platform. The Petri nets [15] are known to have the adequate expressive power to describe behaviour (control structure, internal actions and communication) of distributed and parallel systems. Moreover, they provide several advantages such as a well-founded semantics, an executable specification, and a mean to enable simulations and proofs of behavioural properties.

This whole behaviour is made of four interconnected sub-nets:

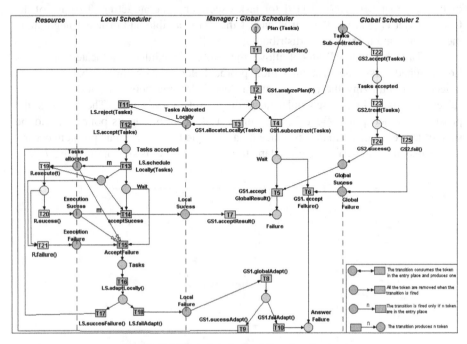

Fig. 8. Petri Net global scheduling functioning

— The Manager is a Global Scheduler in charge of a given user application made of several coordinated tasks (Plan of tasks). The Manager's part of the net is composed of transitions T1 to T10. Once the Manager accepts a plan (T1), it decomposes it into sub-plans (T2) and allocates them to a Local Scheduler (T3) and/or sub-contracts them to another Global Scheduler (T4). If the sub-plan is allocated to the Local Scheduler, the result could be a success stored by the transition T7 in the "Answer Success" place. On the contrary, if the Local Scheduler fails, a global adaptation is performed (T8). This adaptation could fail, and the final result is put in the "Answer failure" place, or could succeed and then a new loop is started by putting the sub-plan in the "Plan Accepted" place in order to schedule it (T1) again and so on.

— The Local Scheduler and resource's part of the net, made up of the transitions T11 to T21, describes the behaviour of a Local Scheduler (LS) and a resource in a local domain. The LS could accept (T12) or reject (T11) an allocated sub-plan. If the sub-plan is rejected, the Manager will try to allocate it again to another LS

(T3) or to sub-contract it to one of its acquaintances (T4). If the LS accepts to treat the sub-plan (T12), the LS schedules it in the local domain (T13). It consists in allocating a resource to each task of the sub-plan and then the execution by the resources could take place. The execution on resources could succeed. If the execution of all the tasks succeeds, the result is put in the "Local Success" place. In any other case, the tasks execution is interrupted and a local adaptation is performed (T16). If the local adaptation succeeds, a new loop is repeated.

— The right part of the net, made up of the transitions T22 to T24, is a view on the behaviour of a Global Scheduler acquaintance (called GS2) playing the role of a sub-contractor for a GS. After the reception of a sub-plan ("sub-plan sub-contracted") place, GS2 will treat it (T22). T22 is an abstract transition which could be expanded to be compliant with any Global Scheduler behaviour (Manager's net). The treatment could succeed and be put in the "Global success place" or could fail and be put in the "Global Failure" place.

It is important to remark that the transitions T3 and T4 are also abstracted transitions corresponding respectively to the tasks allocation and to the contract net protocols as described in section 2.2. The Global Plan (T2) represents the internal behaviour of the Global Scheduler.

3   Typology of Adaptive Grid Scheduling

The Grid environment can be disturbed by various events: resources arrival/departure, resource crashes, priority task arrival, overloads or tasks competition for a resource, etc. The adaptation consists on a set of actions to undertake to maintain or improve a nominal functioning in the case of perturbations. These perturbations are mainly related to resources while the adaptation can relate to various Grid entities described in the conceptual model previously presented (figure 2).

In this section, we define a typology of disturbance events and a typology of adaptations.

3.1   Typology of Disturbance Events

— *Resources entrance/departure.* The Grid environment is an open environment that authorizes the resources to join or leave the Grid freely which can disturb the tasks scheduling and execution. The arrival of new resources disturbs the Grid organization at first but it can be useful to optimize and improve grid performances.

— *Prior tasks arrival.* A task submission (higher priority, more profitable, local, etc.) could modify the order of tasks execution and disturbs tasks allocation.

— *Resource crashes.* It corresponds to the situation where a resource is temporarily unavailable. This problem can occur on a storage device necessary for input data retrieval, on a compute node on which the task is running or on the networks connection (network congestion, overload, etc.). Resource crash can deteriorate the offered quality of service and has many consequences that we detect using our model (task state change, resource state change, organization restructuring, etc).

— *Resource overload.* It is mainly due to applications competition for resources at the execution time. In fact, tasks could be carried out in parallel on the same resource or waiting for resource availability. However a task could act in an unpredicted way by consuming more resource or time and thus hinders the execution of other competing tasks.

3.2   Typology of the Adaptations

We introduced in the previous section the disturbing events. Here we identify the possible types of actions. The types of adaptation are different depending on several criteria: *temporal* (reactive, pro-active), *dynamics* (static or emergent), the *adaptation objects* (plan, process, resources organization, scheduling policy, etc.).

3.2.1 Reactive/Pro-active Adaptation

— *Reactive adaptation.* It assumes a set of corrective actions that are executed as soon as the scheduler detects a disturbing event.
— *Pro-active adaptation.* It assumes a permanent monitoring of the environment which allows to the scheduler to act in an opportunist way. For example, if a resource were requested a long time, it would be judicious to reduce its load. If two resources function in an optimal way when they are associated, it could be beneficial to put them in an organization, etc.

3.2.2 Static/Emergent Adaptation

— *Static adaptation.* It corresponds to an adaptation whose rules are predetermined. For example, we quote the following rule: If "disturbing Event = resource breakdown "Then" compute new tasks scheduling plan".
— *Emergent adaptation.* It assumes that the scheduler has the capacity to produce new adaptation rules.

3.2.3 Adaptation Objects

— *Application process adaptation.* In order to illustrate our remarks, we can formally represent an application process with a Petri net graph [15] where tasks correspond to the transitions, the data necessary for tasks execution to the input places and the data produced to the output places of the corresponding transition. The network structure shows the tasks coordination and can describe various control structures: sequence, alternation and parallelism. From these control structures, we can deduce several execution plans. The adaptation of the application process consists in giving a higher priority to a plan compared to another, removing certain plans that are not realizable in the current context, etc. These adaptations result in the modification of the corresponding network structure of the application process.

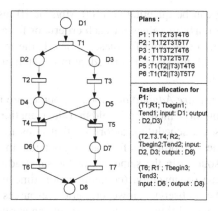

Fig. 9. Application process modelling by a Petri Net

Figure 9 shows an application execution modelled by a Petri net. This application is made up of seven tasks, T1 to T7. The network structure expresses the precedence constraints between tasks. For example, the tasks T2 and T3 need respectively the input data D2 and D3 and can be carried out in parallel. The tasks T4 and T5 need the input data D4 and D5 and are mutually excluded. T5 and T6 are carried out sequentially. Using the following process, a set of all possible application executions plans (P1, P2, P3, P4, P5 and P6) is derived. Scheduling will consist in choosing one of these plans and allocating resources to these tasks. In our example, the plan P1 is selected, and its tasks are assigned to the resources R2 and R1 by specifying the start time, the end time, the input and the output data. We notice that tasks T2, T3 and T4 are associated to the resource R2 to be executed on.

— *Rescheduling*. In the case of disturbances, the tasks to resources mapping must be revised. As a result, a new scheduling plan is computed and carried out by the GS. This type of adaptation requires that the Grid infrastructure offers mechanisms for tasks migration, checkpointing, etc. Most of the adaptive scheduling systems apply this type of adaptation [4], [20], [21].

— *Scheduling policy adaptation*. The Global Scheduler uses scheduling policies which can be fixed or variable and dependent on the resources and their states. The adaptation consists here in modifying this policy when a disturbance occurs. For example, we can quote the change of the use of static scheduling algorithm that uses performance prediction information, to a dynamic algorithm in order to balance the results of the static one and take into consideration the current resources states. Another example is when following the prior tasks arrival, the GS decides to broadcast a call for bid to find the best offer (related to the quality of service, cost, execution time, etc.) instead of using a traditional push or pull policy.

— *Quality of service adaptation*. It assumes at first defining the criteria of quality of service (economic profit, resource utilisation, execution time, application cost execution) which can be quantitative (100 Mb/s or 10 ms) or qualitative (high, average, weak), a range of values for each criteria or for a combinations of these criteria and a certain threshold for these values. If the monitored value exceeds the defined threshold, the system must adapt to maintain it or negotiate to release this threshold [22].

— *Organisational structure Adaptation.* Some resources are not isolated but belong to an organization in which they hold a role (function or type of service). Kreaseck [23] uses a scheduling algorithm that aims to search for a group of candidate machines in order to find an application to a group of resources mapping. The resources in the same site or the same administrative domain are pooled in the same organization (communication time in each subset is lower than between the subsets). However, grid resources can have sufficient intelligence to form coalitions in order to execute an application collectively that the resource could not carry out alone. In this context, the adaptation consists in reorganizing these organisational structures (resources adding or removal, roles modification, etc.). Within the same organization, groups of resources can be formed on the basis of the observation of their collective performance. For example, one can notice that the resources R and S joined together carry out in a more optimal way the application A than the resources S and U, that can lead the GS to give a preference to the first group of resources when the application A will be executed. In the case of coalitions, the resource takes the initiative of modifying its dealings according to the last co-operations and the results obtained with each one of its dealings. This can be managed by reputations mechanisms as the mechanisms used in project CONISE-G [24].

4   An Organizational Model for Grid Scheduling

We have introduced in the previous section a typology of perturbation events and actions that could be undertaken in order to adapt. In our work, we choose to focus on adaptation based on an organizational approach. For that purpose, we have modelled the grid scheduling system with an organizational view. In fact, this approach offers several advantages for modelling the grid system in general and the grid scheduling in particular. In fact, the design and implementation of a grid scheduling system is a difficult task due to the grid environment constraints (resources distribution, autonomy, performance variation, etc.). The organizational perspective constitutes a design support since it makes it possible to apprehend and to structure a multi-agents system (MAS) through various roles and their interactions. Moreover, the grid scheduling system operates through the co-operation of many interacting subsystems. The organizational perspective structures the MAS execution since it defines, through the attribution of roles to the agents, behavioural and interaction rules to which the agents must conform. Finally, the grid environment is an open and dynamic environment. The organizational perspective allows the design of open systems with heterogeneous components where agent internal architectures are not specified.

The system can thus be adapted to an increased problem size by adding new roles and groups, and this does not affect the functionality of the other agents.

The remainder of this section first presents the Agent Group Role (AGR) meta-model of Ferber [11]. In fact, our organizational model for the grid scheduling is based on the AGR meta-model which is appropriate to the grid context. Then, we present how we apply this meta-model in order to structure and organize our grid scheduling system.

4.1   The AGR Meta Model

In this section, we describe the AGR Meta Model [11] which core concepts are Agent, Group and Role.

- An *agent* is defined as an active communicating entity that can participate in several communities (groups) in parallel. It can play one or several roles corresponding to its activities or interactions in each group. No constraint is placed on the internal structure of the agent.
- A *group* is defined as a set of agents. Agents can belong to different groups. The communication between agents is possible only if they belong to the same group space. Communication between two groups is made by agents that belong to both.
- A *role* is a representation of an agent function. Agent may play one or several role within a group.

AGR has several advantages useful in our context: it eases the *modularity* through the organization of tasks carried out by agents and their interactions in logical entities: working group, organizational unit or private space of conversations. The organization of the tasks carried out by the agents in logical entities enhances the *security* of applications. In fact, in the AGR model, security mechanisms such as entrance protocol, authorizations and permissions can be integrated in order to define how an agent can enter, leave, and behave inside the organization. As a result, agents not belonging to a group cannot listen or join freely in a conversation in this group. This allows secure interactions inside and between groups. Moreover, at the role level, describing norms such as obligations, permissions, interdictions, etc. allows to prevent unauthorized actions to be executed by agents. Also, AGR allows the design of *open* systems such as grid. The system can thus be adapted to an increased problem size by adding new roles and groups and this does not affect the functionality of the other agents. The system can also be designed with *heterogeneous* components where agent internal architectures are not specified. Finally, the *reusability* is facilitated since the organizational approach identifies, expresses and makes available recurring interaction patterns which become reusable.

4.2   An AGR Organizational Model for Grid Scheduling

In this section, we describe our organizational model which identifies the actors implied in our system, the links that exist between them, and the protocols that rule their interactions.

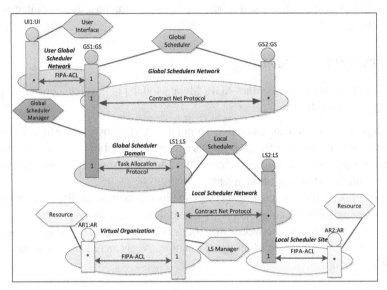

Fig. 10. Organizational multi-agents based grid architecture

Our organization model (see figure 10) organizes the architecture' agents introduced in section 2.2 around the following components:

— Four types of agents represented by a candle that are: the User Interface Agent, the *Global Scheduler Agent*, the *Local Scheduler Agent* and the *Resource Agent*. The multiplicity of the agents participating to a group is represented by a star inside the candle.

— Six types of groups represented by an ellipse that are: the Users *Global Scheduler Network*, the *Global Schedulers Network*, the *Global Schedulers Domain*, the *Local Scheduler Site*, the *Local Schedulers Network* and the *Virtual Organization*.

— Six types of roles since each agent can plan different roles in different groups. The defined roles are: the *User Interface Role*, the *Global Scheduler Manager Role* that is a GS playing the role of tasks manager, the *Global Scheduler Contractor Role* that is a GS acquaintance responsible of treating subcontracted tasks, the *Local Scheduler Role,* the *Resource Role* and the *Local Scheduler Manager Role*. The Local Scheduler Manager is responsible for managing a virtual organization. In fact, as we previously mentioned the GS allocates tasks on selected resources that will form a virtual organization made by multiple distributed resources for application tasks execution. When a virtual organization is formed, an agent playing the role Local Scheduler Manager is launched by the

Global Scheduler in order to orchestrate the different tasks in the virtual organization and to adapt the organization functioning in case of disturbances. The adaptation includes recruiting a new resource to participate to the organization, excluding resources that slowdown the execution performance, rescheduling tasks in the organization, etc. Role is represented as a hexagon and a line links this hexagon to agents.

The communication between agents in different groups follows the FIPA-ACL language that supports high level communication between the different grid components (see the section 2.2 for more details).

Let us detail how does each group operates.

1. The *User Global Scheduler (GS) Network* Group allows user interface agents to ask for application execution, to submit user applications and to collect the execution results.

2. The *Global Scheduler (GS) Network* Group. This group allows the GS to cooperate in order to allocate application's tasks efficiently when resources under a Global Scheduler control are not available or in the case of disturbance. In this context, Global Scheduler (GS) agents cooperate in order to sub-contract the execution of application's tasks using the *Contract Net Protocol* [17] (see the section 2.2 for more details).

3. The *Global Scheduler Domain* Group is composed by the Global Scheduler and the Local Scheduler agents at its disposal. This group allows the Global Scheduler to allocate the application's tasks to the adequate Local Scheduler to be executed on its local resources. The allocation is based on the Global Scheduler objective, resources states in the local domains and tasks requirements. Moreover, this group allows the Local scheduler to send execution results, resources states and to inform its Global Scheduler if some problem occurs on resources to allocate the failed tasks on another resource.

4. The *Local Scheduler Site* Group is composed by the Local Scheduler (LS) and the resource at its disposal (in the local site). This group allows to the Local Scheduler to allocate the submitted tasks on the local resources, to collect the resources information (availabilities, breakdown, task execution progress, etc.), the execution results, etc.

5. The *Virtual Organization* Group is composed by selected resources for the application execution and by Local Scheduler agents. In fact, in each formed virtual organization, a Local Scheduler Manager coordinates the functioning of resources inside the organization, monitors their execution and undertakes adequate action to adapt in case of perturbation (resource breakdown, resource overload, etc.). In case of perturbation, adaptation consists in reorganizing the virtual organization by adding a new better resource using the contract net protocol, removing an overloaded resource, replacing resource by another, etc. The adaptation results in the modification of the structure of the virtual organization.

6. The *Local Scheduler Network* Group is made by local scheduler agents in the same virtual organization. This group allows to the Local Scheduler Manager to cooperate efficiently with other Local Scheduler in order to add new resources for tasks execution in case of perturbation and when resources at the LS Manager disposal are not available, busy, broken, etc. The LS cooperate using the Contract Net Protocol described above.

In the introduced groups, adaptation is made thanks to the use of high-level protocols, like contract net protocol that eases task allocation even if the grid environment evolves. The use of interaction protocols allows replacing one resource by another, sending tasks to be executed by acquaintance, etc. Here, the structure of the organization remains the same.

5   The Organizational Grid Scheduling Simulator "OGSSim" with the AGR Meta Model

We have implemented an organizational grid scheduling simulator based on the AGR meta-model using Java and the Madkit [12] platform that (or which) integrates the AGR meta-model.

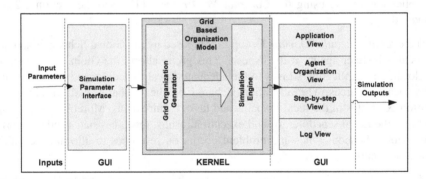

Fig. 11. Multi-agents based grid simulator

Our simulator offers the following services:

— It allows users to enter parameters for the simulations,
— It simulates the applications' tasks scheduling and their execution step by step,
— It finally outputs the statistical data of the performance metrics (execution time, number of messages, etc.) on a log file.

The structure of our simulator is illustrated in figure 11. It includes an *input GUI,* a *kernel* and an *output GUI.*

Figure 12 shows some screenshots of our GUI. We distinguish between two types of GUIs: the *input GUI* (left side Fig.12) and the *output GUI.*

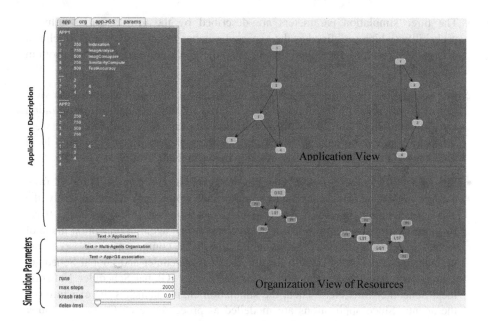

Fig. 12. The simulator's GUI

- The *input GUI* allows the user 1) to define the *grid specification*, the *application description*, the *simulation parameters* and 2) to display this data in a graphical form.

1. The first service concern defining the simulated environment and more precisely:

- The *grid specification* described by:

  — The *resources* (number, identifier) with their computing capacities rated as MIPS (Million Instructions Per sec) and theirs skills (types of tasks they are able to perform).
  — The *physical resource organization* (in a domain, site, etc.) and thus the number of Local Scheduler and Global Scheduler in the simulated system. It is described thanks to the second slot called "org" in our input GUI (Fig. 12).

- The applications are described by their tasks, their coordination and their submission time. For each task, we provide the number of instructions, the type and tasks' coordination process (preceding constraints). The top left side of Figure 12 shows an example of two applications submitted to the grid. The first one is composed of five tasks and three preceding constraints (the constraints line 1 means that tasks 1 must be performed before task 2 and 4). Each task has a certain number of instructions and a type. The second application is composed of four coordinated tasks, etc.

- The three simulation parameters are described by the grid resource failure's probability (which is defined as the expected number of failures per step which is a discrete unit of time), the number of runs and the maximum execution step (bottom left Fig 12).

2. The second service concerns *input GUI* allows also displaying an application view (top right figure 12) of the applications entered by the user and an organization view (bottom right Fig.12) of the resources. For example, the organization view shows two directed graphs in which vertex represent the Global Scheduler, local scheduler and resource an edge the relation between them.

- The *KERNEL* has two roles. It generates the agent based organization model from a user's inputs and it launches the execution of the simulations. More precisely, the grid organization generator (figure 11) transforms grid information (Grid Resources, Global scheduler, Local Scheduler) into an agent-organization made of Agent-Group-Role model. For example, the grid components (resource and organization) are converted into Resource, Local Scheduler and Global Scheduler agents, each agent belong to one or several groups (Global Scheduler Network, local Scheduler Site, etc.). Then, the simulation engine launches the execution of the agents. Since applications are modelled as processes, each Global Scheduler agent behaves as a workflow engine (or orchestrator). Finally, when the simulation ends, results are displayed to the user on the *GUI output*.

- The *output GUI* displays the simulations results on a log file. The structure of this file is a tuple of the following form < Run Number, ApplicationID, Application Start time, application End Time, TaskID, Task Start Time, Task End Time, Number of exchanged Messages>.

6   Evaluation of Our Model

6.1   Performance Metrics

In this section we introduce the performance criteria used in our approach and the evaluation of our organizational model for grid scheduling based on these criteria.

Our study covers two aspects: the *structural* and the *statistical* aspect:

- The *structural level* concerns the communication topology of the organization and is based on the Grossi's structural measures [13] that are described below. These measures allow to proof organizational qualities such as the flexibility (the capacity of the organization to adapt in a flexible way to changing circumstances), the robustness (how stable the organization is in the case of anticipated risks) and the efficiently (refers to the amount of resources used by the organization to perform its tasks) by using relations holding between roles. In our case, we will evaluate this qualities and proof that our model is sufficiently robust and the flexible. These two qualities are the most important in our grid scheduling context targeted to execute users' applications efficiently while taking into account the dynamicity and complex interactions among the different components.

- The *statistical level* concerns the evaluation of our organization model performances at run time. The used metrics are intended to observe the functioning of our organizational based grid in the case of disturbances to proof its robustness. To do that two metrics are observed related to the grid and to multi-agent system (MAS). The robustness of a system is defined as the ability of the system to maintain its functioning in highly dynamic environments such as the grid.

  - *The grid metric* corresponds to the application execution time in a scenario without perturbation and in a scenario with perturbations.
  - *The MAS metric* corresponds to the number of exchanged messages that is commonly utilized for evaluating multi-agents system in a scenario without perturbation and in a scenario with perturbations [14].

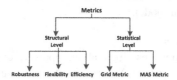

Fig. 13. Performance metrics

Let' us details the ***Structural level*** based on Grossi's [13] measures. Evaluating the organizational structure property involves three steps:

3. The first step consists of building a role graph of the organization based on the possible relations between two roles. To do that, Grossi introduces three dimensions characterizing relations:

- The *power structure* defines the task delegation patterns (existence of power link between agent *a* and *b* means that every delegation of tasks from agent *a* to agent *b* ends in creating an obligation to agent *b*),
- The *coordination structure* concerns the flow of knowledge within the organization.
- The *control structure* deals with the task recovery functions, that means that an agent "*a*" that controls one another "*b*" has to monitor its activity and possibly take over the tasks which agent *b* has not accomplished.

4. Secondly, for *each dimension* (power, coordination, control), a set of concepts and equations from the graph theory are applied in order to measure specific property of organizational structures (OS). Grossi defines the following concepts:

- The *connectedness* of an OS shows the connection degree between roles. The more this degree is high, the more the structure can be divided into fragments.
- The *economy* of an OS expresses how to keep the structure connected while minimizing the number of links between roles (for example redundant links must be avoided).
- The *univocity* allows us to have an idea on the degree of ambiguity in the organization structure. For example stating that an agent "a" controls an agent "b"

and in the same way, agent "b" controls agent "a" generates some ambiguity. It is a ratio between the number of roles which have exactly one link in the same structural dimension and the total number of roles.

5. Finally, these previous concepts are compared with optimum values defined by the author (See [13] for more details) in order to measure the qualities (robustness, the flexibility and the efficiency) of the organizational structure. In our context, we will calculate these measures and compare them to the optimum values to evaluate the qualities of our model.

6.2   Evaluation of Our Organizational Model

In this section we describe the structural and statistical evaluation (experimentation results) of our organizational model.

6.2.1   Structural Evaluation

According to the dimensions described above (power, coordination, control), we generate the role graph (figure 14) corresponding to our organizational model. Our graph includes the roles described in section 4.2.

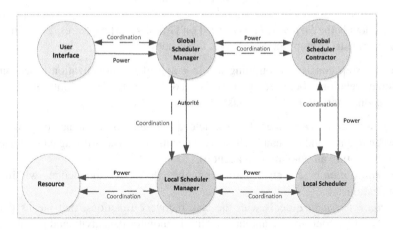

Fig. 14. The roles graph

Our graph includes only the power and coordination dimensions represented as edges:

- *Power* expresses the relation which can occur between an agent "delegator" and the one who receives the delegation "delegatee". The delegation is a mechanism allowing the delegator to assign tasks to the delegatee. In our context, the user interface delegates tasks to the Global Scheduler Manager. The Global Schedulers (Manager and contractor) can delegate tasks to the Local Schedulers under their control and to their GS acquaintances. The Local Schedulers (including the Manager) has a power relation with the Resources under their control and with their Local Scheduler acquaintances in the same virtual organization.

- *Coordination*: In our graph, the coordination is a symmetric relation. It consists in information exchange or protocols-based interaction between roles. It concerns the same couple of roles as the power relation.

The *Control* relation is not present in our organization, since we do not model the evaluation of agent's behaviour by their acquaintances. In future work, we intend to refine our model to include this dimension in order to improve the adaptation process. This assumes to model the continuous monitoring of the agent work to check whether agents are doing their tasks as agreed.

Following this role graph and the set of equations introduced in [13], we compute the three following organizational qualities: the robustness, the flexibility and the efficiency as shown in the following tables:

Table 1. Structural measures of robustness

Optimum Values	Economy $_{Coord}$	0	Univocity $_{Power}$	0	Connect$_{Coord}$	1
Obtained Values		0,72		0.2		1

Table 2. Structural measures of flexibility

Optimum Values	Connect$_{Power}$	0	Connect$_{Coord}$	1	Economy$_{Coord}$	0
Obtained Values		0,64		1		0,72

Table 3: Structural measures of efficiency

Optimum Values	Economy $_{Power}$	1	Economy $_{Control}$	1	Economy $_{Coord}$	1
Obtained Values		0,96		1		0,72

Comparing our obtained results to the optimum ones, we can state that our system is sufficiently *robust* and *efficient* but not enough *flexible*. Lets' us explain these results:

- *Robustness* of an organization requires a coordination structure highly connected and weakly economic and a power structure weakly univocal (see Table1). In our organization, roles are well connected (Connectedness$_{Coordination}$ value equal to the optimum) while avoiding ambiguities (Univocity$_{Power}$ value approaches the optimum). An optimal robustness would require a complete connectivity between all nodes. Due to the privacy and local management policy, this complete connectivity is not possible in a grid context. For example each resource in a specific domain has its own management policy and doesn't authorize the Global Scheduler to interact directly with it.
- *Flexibility* of an organization is related to the ability to easily adapt. So, in order to enhance the flexibility, a low degree of connectedness for the power dimension and a high degree for the coordination dimension are needed. Considering the connectivity aspect, while the power relation restrains the flexibility (the delegation pattern imposes constraints on the interactions), the coordination one eases the interactions within the organization: we reach a

trade-off (see Table 2). Considering the economy aspect, we have an average value which corresponds to a moderate flexibility. To obtain a lower value of the Economy$_{Coordination}$, for more flexibility, we should increase the redundancy of the coordination relation which is as previously noticed incompatible with the privacy constraint of the grid context.

- *Efficiency* of an organization can be obtained if it is economic (value equal to 1) in all the dimensions. In our case, the roles are well connected while avoiding redundant ones and we approach the optimal efficiency. The value of the Economy$_{Control}$ is equal to the optimum one since we don't take into consideration this dimension. The Economy$_{power}$ is quite optimal while the Economy$_{coordination}$ has an average value.

It is important to notice that the framework of Grossi [13], used for analysing organizational structures, does not take into consideration the number of roles' instances and the links among them. However, in our context, introducing the roles' cardinalities could be more significant and realistic. For example, a broken resource doesn't imply that the resource role dysfunctions, but this specific instance is out of work and could be replaced by one another.

Also, obviously, designing an organization that maximizes simultaneously the three qualities cannot be reached. For example, as we explained above, the more the organization' nodes are well and directly connected, the more the organization is robust and flexible and the less it is efficient.

Since our organization is devoted to the grid scheduling system, we can claim that the qualities obtained (sufficiently robust and efficient) of our organization are adequate for this context. Decreasing the number of power relations can improve the flexibility but it will decrease the robustness.

As our context is highly dynamic and characterized by a frequent resources performances fluctuations and crashes, we are also interested in the capacity of our organization to be adaptive. The adaptation is the ability of the system to react against the environment changes. The adaptability cannot be measured using the previous framework. In the next section, we measure the quality of our model experimentally by simulations and more precisely its adaptability. For this purpose, we compare the nominal functioning of our system and its functioning with disturbances.

6.2.2 Statistical Evaluation

In this section, we describe our experiments and the obtained results.

Our experimental system is configured with twelve agents, illustrated by the figure 15. The Global Schedulers are represented with a square (GS1, GS2), the Local Schedulers are represented with an ellipse and resource with a rounded rectangle (including the CPU speed). The following grid specification is simulated and organized according to our proposed AGR model described in section 4.2.

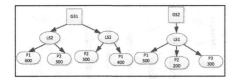

Fig. 15. Case study: agent's organization   **Fig. 16.** Case study: application description

The goal of these simulations is to demonstrate the feasibility (figure 17) and the efficiency (figures 18, 19, 20, 21 and 22) of our model and more precisely of its adaptation.

To demonstrate *the feasibility of our proposition*, we computed the number of successful executions of applications in the case of adaptation comparing to the case with no adaptation. To do that, we considered four applications (figure 16 shows an example of an application description) and vary the resources' failure probability. We measured the percentage of successful runs while varying this failure probability. The total number of run is fixed to 100.

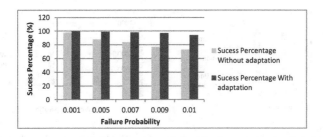

Fig. 17. Execution success percentage

The results show the efficiency of our adaptation model since we obtained better percentage of successful executions. We can however notice that we do not reach 100% of success since even with adaptation we can meet situations where some tasks could not be executed by lack of specific resources.

To demonstrate the *efficiency of the adaptation* of our model, we made three experiments. In the first one, we considered only one application and vary the resource's failure probability. Figure 18 compares the application execution time with and without disturbances. The results show that the adaptation phase is not costly (at the maximum 27.05 % of the application execution time). Figure 19 shows how the number of communications increases as the failure probability raises. This evolution is due to the fact that adaptation is mainly based on interactions.

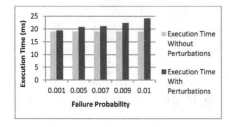
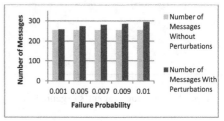

Fig. 18. Comparison of application execution time **Fig. 19.** Comparison of number of messages

The second experiment varied the number of concurrent applications executed by the system with a fixed resource's failure probability (0.01). Figure 20 shows again that the adaptation time remains low even if the number of the number of applications increases (20.8% of the application execution time). Figure 21 shows that the number of messages increases with the number of applications in both cases: with or without perturbations. In case of perturbations, we noticed a higher but reasonable number of communications due to the adaptation phase.

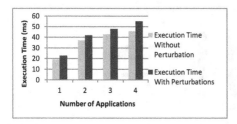
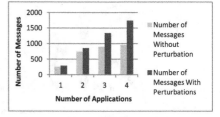

Fig. 20. Comparison of application execution time **Fig. 21.** Comparison of number of messages

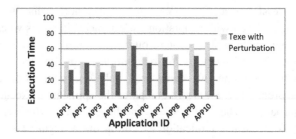

Fig. 22. Comparison of application execution time

In the third experiment, we increased the number of processors to fifteen, the number of Local Scheduler to forty and the number of Global Scheduler to ten. The aim is to check the system performance scalability. We considered ten applications. Each application arrived to the system every unit of time. We fixed the resource's failure probability to 0.01. Each result is the average value that is derived from 100 simulation experiments. Figure 22 compares the execution time for each application in the case of perturbations and the case where there is no perturbation. Figure 22 shows again that the adaptation time remains low even if the number of applications increases (the average is 20.86% of the application execution time).

7   Related Work

The problem of adaptation in grid computing has been investigated in many works in the literature. As a result, adaptive scheduling systems have already been proposed [20][23][25][26], but these systems show some restrictions.

The condor system [27] offers rescheduling and checkpointing mechanisms allowing the application to be restarted from checkpoints. Rescheduling consists in migrating the impacted application (or portion of application) to another resource when a perturbation occurs. Buyya et al. propose the Nimod/G [28] and Grace systems [29] that implement an economics model for grid resource management and scheduling. Marketing concepts such as commodity market, posted price and bargaining modelling are proposed for resource trading and scheduling. Moreover, the Nimrod/G and Grace systems incorporate schedule adaptation in order to meet several user QoS (quality of service) requirements (deadline or budget), and also to adapt to resource availability and performances. It is interesting to note that our proposition can be used to build a system like Nimrod/G since we use the contract net protocol for resource trading. However, in our system the scheduling component decides which tasks are to be executed at which site based on a certain cost function (in the current implementation, the global scheduler objective is to minimize the application makespan). Unlike our system, the Nimrod/G system supports user-defined deadline and budget constraints for application scheduling and manages the supply and demand of resources based on resources prices. Moreover, the introduced systems (condor and Nimrod/G) implement a centralized strategy. The scheduling decision is made by one centralized component. This approach has the advantage of easier implementation, but suffers from the lack of scalability, the possibility of generating a bottleneck and fault tolerance since the failure of the scheduling component will result in the entire system failure.

The system DIET [30] and Legion [31] implement hierarchical scheduling combined with rescheduling techniques. In the hierarchical scheduling [32], there is a central scheduler that distributes applications tasks to multiple lower-level sub schedulers. Unfortunately, even if the hierarchical strategy offers a better scalability than the centralized one, the failure of the central scheduler will also result in the entire system failure.

Unlike the hierarchical strategy, in the decentralized architecture there are multiple schedulers without a central one. Each scheduler has the responsibility to carry out its own portion of scheduling and can communicate with other schedulers to allocate tasks to another one with lower load. This naturally led us to choose a decentralized strategy in order to implement an adaptive grid scheduling more scalable than the centralized and hierarchical scheduling and offering naturally fault tolerance against failures.

In the decentralized scheduling, the scheduling components can work independently or cooperatively. In the independent scheduling, each scheduler acts as an autonomous entity and takes scheduling decision regardless of the effects of the decision on the rest of the system. The systems GraDs [4][20][33] and AppLeS [5] are two well-known adaptive grid scheduling systems that integrate rescheduling techniques. Both implement a non-cooperative strategy. In fact, the scheduler is tightly integrated in the application and optimizes its private individual objectives. As a result, these systems are not easily maintainable and are not easily applied to other types of applications. Here, a significant advantage of our system compared to the AppLeS and GraDs systems is the reusability of the scheduling architecture for various applications since the scheduling component is decoupled from the application to execute on the grid.

Our framework implements a decentralized and cooperative scheduling. In the cooperative scheduling, each scheduler makes scheduling decisions considering the other schedulers in order to achieve a global goal. Multi-agent systems (MAS) have been used with success for managing the grid environment in a decentralized and cooperative way. In fact, the benefits of MAS are widely recognized in the literature [9]. This paradigm is one of the promising approaches due to the characteristics of the agent like autonomy, proactivity, mobility and adaptability allowing to take naturally into consideration the complexity of the grid context. Moreover, MAS allows the design of software components that exhibit social capacities that could be used with benefits for managing the interaction among the grid components. As a consequence, a great number of systems applying the agent paradigm have been proposed such as the AgentScape [34], the AGEGC [35], the CONOISE [24] projects, the AGRD_NFRP [36] system, etc. Unfortunately, most of these systems neglect the adaptation process. However, we must mention the system ARMS [37] that implements an adaptive scheduling. The system ARMS [37] is an agent based resource management that uses a hierarchy of homogenous agents for scheduling grid applications efficiently. The system ARMS focuses on the dynamicity of the resource performances and the scalability of the grid. However, this work suffers from a lack of flexibility due to its hierarchical structure, which makes it unable to perform well in the rapidly changing environments of the grid. Finally, in [38], a MAS is proposed in which a set of event-condition-action rules is applied to guide the adaptation process. The adaptation is made either by reassigning the failed task to another agent for completion or by asking the user to relax its constraints. The advantage of this approach is its extensibility since adaptation rules can be added easily to the rule base. The limit is that evaluation and rescheduling are repeated iteratively at runtime which can incur a certain overhead and decrease the system performance. Moreover, all these systems impose an internal structure to the agents and do not consider their organization and regulation to reach the system objective. On the other hand, our system implements an organizational perspective allowing to structure and regulate the scheduling system and to improve its efficiency.

A last limitation that we can address to the systems presented above (GraDs, AppeLs, Nimor/G, etc.) is that the adaptation is limited to rescheduling. When a perturbation occurs on a resource, the adaptation consists in migrating the failed tasks to another resource to adjust to changes in resource availabilities. Obviously, as described in section 3.2 and in our previous work [38], other types of adaptation can be investigated such as application process adaptation, grid resource reorganization, quality of service adaptation, etc. In our work, adaptation is based on rescheduling, on resource reorganizations and on the use of high level interactions protocols. In fact, rescheduling is carried out by the local and global scheduler agents (see section 4.2). Reorganization is made by the virtual organization scheduler, in case of disturbances. Such restructuring includes integrating a new more suitable resource, excluding a resource slowing the execution, etc. Adaptation is made thanks to the use of the contract net protocol that eases task allocation even if the grid environment evolves (call for proposal to replace one resource by another, to send tasks to be executed by acquaintance, etc.). As a result the offered mechanisms ease the adaptation and make it more efficient.

8   Conclusion

Grid computing is known to be a heterogeneous, distributed and dynamic environment. In order to take fully advantages of computational grid power, grid scheduling must take into consideration the grid constraints (resource availability variation, prior task submission, breakout, etc.) and be adaptive. In this work, we proposed a conceptualization of the grid environment used to define a typology of the adaptation. The defined typology introduces the perturbing events and the possible adaptation actions. Adaptation can be related to the environment and to the software components with their interactions and organizations. In our work, we choose to focus on the adaptation according to an organizational view. More precisely, we introduce a framework for designing an evaluating a multi-agent organizational model for an adaptive grid scheduling. The framework includes an Agent Group Role (AGR) organizational model to describe and implement an adaptive grid scheduling system, an organizational grid scheduling simulator "OGSSim" and performance criteria to evaluate the proposed model. The proposed model has been validated: it has been implemented and its design and performance has been discussed. Following the Grossi's framework we have evaluated its conceptual aspect and shown that it is sufficiently robust and efficient. According to our implementation, we have shown the feasibility and efficiency of our approach and more precisely measured the adaptability of our organizational model. In addition, the proposed system presents the following qualitative advantages:

Easy Design and Implementation of the Grid Scheduling System. The design of such system is a difficult task due to the constraints of the environment (complex interactions, autonomous and heterogeneous components, disturbed, etc.). The organizational perspective constitutes a design support and makes it possible to structure the overall functioning through the attribution of roles and interaction rules to which the agents must conform.

Openness. The grid is an open environment allowing the resources to enter and leave at any moment. The use of an organizational perspective allows the design of open systems such as grid since organizations are open structures allowing 1) to agents playing specific roles to enter groups without limits, 2) to add new roles and groups and this does not affect the functionality of the other agents.

Dynamicity. The grid is a dynamic environment were resources exhibit variable performances due to the openness of the grid, possible breakdowns, tasks executions concurrence, etc. The organizational model is well-adapted for such context since organizations are active entities that can be able to reorganize dynamically and to adapt to the environment changes.

Security. Security is a key requirement to ensure secure application executions on the grid and involves protection of data exchanges between tasks, controlling access to computers, data and other resources, etc. The organizational perspective allows the design and implementation of secure grid scheduling architecture. In fact, our proposed model organizes the grid components in groups where security policies such as protocols entrance (authorization and authentication procedures) can be integrated to restrict access to only authorized entities and keep the undesirable ones out of a group. Communications and data exchanges inside and between groups are also protected against malicious attacks since our agents are organized in private spaces and agents that do not belong to a group cannot listen or freely integrate in a conversation in this group, etc.

Reusable Framework. Our work is reusable and could be considered as a first step towards a framework for designing and testing grid organizations. Indeed, the designers could use our system to model their own organizations, measure their qualities (robustness, flexibility, efficiency), tune their organizations in order to fit with their requirements, validate these qualities and finally simulate their functioning.

This work opens a number of issues for future research.

The first issues we are currently investigating are related to the *structural evaluation* of our organization:

Firstly, how to add the *control dimension* in our model since we have only implemented the power and coordination dimensions and what are the effects on the qualities of our organization?

Secondly, we do believe that a refinement of the theoretical evaluation is needed. Indeed Grossi framework doesn't take into account the occurrences of roles involved in the modelled organizations. However, taking into account the roles cardinalities and the relations occurrences will certainly influence the evaluation of our model.

Regarding future work, since we have only proven the feasibility of our model, we intend to compare our model with existing scheduler such as the system ARMS that implements adaptive scheduling based on hierarchical multi-agents system. More precisely, we need to compare the adaptiveness and the efficiency of these systems, in particular, interaction overhead, robustness and scalability in case of disturbances.

Finally, we need to consider more applications and possibly in a real grid environment.

Acknowledgements. We thank the anonymous referees for critically reading the manuscript and making several useful remarks. We also wish to thank our colleague Rui Tramontin from the Federal University of Santa Catarina, Brazil, for kindly reading our paper and offering valuable suggestions.

References

1. Foster, I., Kesselman, C.: The grid: blueprint for a new computing infrastructure. Morgan Kaufmann (2004)
2. Dong, F., Akl, S.G.: Scheduling algorithms for grid computing: State of the art and open problems. School of Computing, Queen's University, Kingston, Ontario (2006)
3. Schopf, J.M.: Ten actions when grid scheduling. International Series in Operations Research and Management Science, 15–24 (2003)
4. Wrzesinska, G., Maassen, J., Bal, H.E.: Self-adaptive applications on the grid. In: Proceedings of the 12th ACM SIGPLAN Symposium on Principles and Practice of Parallel Programming, New York, NY, USA, pp. 121–129 (2007)
5. Berman, F., et al.: Adaptive computing on the grid using AppLeS. IEEE Transactions on Parallel and Distributed Systems 14(4), 369–382 (2003)
6. Berman, F., et al.: New Grid Scheduling and Rescheduling Methods in the GrADS Project. International Journal of Parallel Programming 33(2-3), 209–229 (2005)
7. Foster, I., Kesselman, C., Tuecke, S.: The anatomy of the grid: Enabling scalable virtual organizations. International Journal of High Performance Computing Applications 15(3), 200 (2001)
8. Dignum, V.: The Role of Organization in Agent Systems. Multi-agent Systems: Semantics and Dynamics of Organizational Models. IGI (2009)
9. Foster, I., Kesselman, C., Jennings, N.: Brain Meets Brawn: Why Grid and Agents Need Each Other. In: Proceedings of the Third International Joint Conference on Autonomous Agents and Multi-Agent Systems, pp. 8–15. IEEE Computer Society (2004)
10. Demazeau, Y.: Invited lecture, 1st Ibero-American Workshop on Distributed AI and Multi-Agent Systems (IWDAIMAS 1996), Mexico (1996)
11. Ferber, J., Gutknecht, O., Michel, F.: From agents to organizations: an organizational view of multi-agent systems. Agent-Oriented Software Engineering IV, 443–459 (2003)
12. The MADKIT Agent Platform Architecture, http://www.madkit.org
13. Grossi, D., Dignum, F., Dignum, V., Dastani, M., Royakkers, L.: Structural evaluation of agent organizations. In: Proceedings of the Fifth International Joint Conference on Autonomous Agents and Multiagent Systems, New York, NY, USA, pp. 1110–1112 (2006)
14. Kaddoum, E., Gleizes, M.P., Georgé, J.P., Picard, G.: Characterizing and evaluating problem solving self-* systems. In: Computation World: Future Computing, Service Computation, Cognitive, Adaptive, Content, Patterns, pp. 137–145 (2009)
15. Petri, C.A.: Fundamentals of a Theory of Asynchronous Information Flow, Amsterdam. Presented at the IFIP Congress 62, pp. 386–390 (1962)
16. Fibich, P., Matyska, L., Rudová, H.: Model of grid scheduling problem. In: Exploring Planning and Scheduling for Web Services, Grid and Autonomic Computing, pp. 17–24 (2005)
17. Smith, R.G.: The contract net protocol: High-level communication and control in a distributed problem solver. IEEE Transactions on Computers 100(12), 1104–1113 (2006)
18. Foundation for Intelligent Physicals Agents, http://www.fipa.org

19. Foundation for Intelligent Physicals Agents: FIPA Contract Net Interaction Protocol Specification, http://www.fipa.org/specs/fipa00029/SC00029H.pdf

20. Vadhiyar, S.S., Dongarra, J.J.: Self adaptivity in grid computing. Concurrency and Computation: Practice and Experience 17(2-4), 235–257 (2005)

21. Reed, D.A., Mendes, C.L.: Intelligent Monitoring for Adaptation in Grid Applications. Proceedings of the IEEE 93(2), 426–435 (2005)

22. Iosup, A., et al.: On grid performance evaluation using synthetic workloads. In: Frachtenberg, E., Schwiegelshohn, U. (eds.) JSSPP 2006. LNCS, vol. 4376, pp. 232–255. Springer, Heidelberg (2007)

23. Kreaseck, B., Carter, L., Casanova, H., Ferrante, J.: Autonomous protocols for bandwidth-centric scheduling of independent-task applications. Presented at the 17th International Parallel and Distributed Processing Symposium, IPDPS 2003 (2003)

24. Patel, J., et al.: CONOISE-G: agent-based virtual organisations. In: Proceedings of the Fifth International Joint Conference on Autonomous Agents and Multiagent Systems, New York, NY, USA, pp. 1459–1460 (2006)

25. Buisson, J., André, F., Pazat, J.-L.: Dynamic Adaptation for Grid Computing. In: Sloot, P.M.A., Hoekstra, A.G., Priol, T., Reinefeld, A., Bubak, M. (eds.) EGC 2005. LNCS, vol. 3470, pp. 538–547. Springer, Heidelberg (2005)

26. Therasa, A.L.S., Sumathi, G., Dalya, A.S.: Dynamic Adaptation of Checkpoints and Rescheduling in Grid Computing. International Journal of Computer Applications 2(3), 95–99 (2010)

27. Condor Project Homepage, http://research.cs.wisc.edu/condor/2006

28. Abramson, D., Buyya, R., Giddy, J.: A Computational Economy for Grid Computing and its Implementation in the Nimrod-G Resource Broker. Future Generation Computer Systems (FGCS) Journal 18(8), 1061–1074 (2002)

29. Buyya, R., Abrasmson, D., Venugopal, S.: The Grid Economy. Proceedings of the IEEE 93(3), 698–714 (2005)

30. Caron, E., Desprez, F.: Diet: A scalable toolbox to build network enabled servers on the grid. International Journal of High Performance Computing Applications 20(3), 335 (2006)

31. Chapin, S., Katramatos, D., Karpovich, J., Grimshaw, A.: The legion resource management system. In: Job Scheduling Strategies for Parallel Processing, pp. 162–178 (1999)

32. Cao, J., Spooner, D.P., Jarvis, S.A., Nudd, G.R.: Grid load balancing using intelligent agents. Future Generation Computer Systems 21(1), 135–149 (2005)

33. Dail, H., et al.: Scheduling in the grid application development software project. International Series In Operations Research and Management Science, pp. 73–98 (2003)

34. Wijngaards, N.J.E., Overeinder, B.J., van Steen, M., Brazier, F.M.T.: Supporting internet-scale multi-agent systems. Data & Knowledge Engineering 41, 229–245 (2002)

35. Shi, Z., Huang, H., Luo, J., Lin, F., Zhang, H.: Agent-based grid computing. Applied Mathematical Modelling 30(7), 629–640 (2006)

36. Muthuchelvi, P., Anandha Mala, G.S.: Agent Based Grid Resource Discovery with Negotiated Alternate Solution and Non-Functional Requirement Preferences. Journal of Computer Science (2009), http://www.scipub.org/fulltext/jcs/jcs53191-198.pdf

37. Cao, J., Jarvis, S.A., Saini, S., Kerbyson, D.J., Nudd, G.R.: ARMS: an Agent-based Resource Management System for Grid Computing. Scientific Programming 10(2), 135–148 (2002)

38. Wang, M., Ramamohanarao, K., Chen, J.: Robust Scheduling and Runtime Adaptation of Multi-agent Plan Execution. In: Proceedings of the 2008 IEEE/WIC/ACM International Conference on Web Intelligence and Intelligent Agent Technology, WI-IAT 2008, vol. 2, pp. 366–372. IEEE Computer Society, Washington, DC (2008), http://dx.doi.org/10.1109/WIIAT.2008.136

39. Thabet, I., Hanachi, C., Ghédira, K.: Vers une Architecture de Type Agent BDI pour un Ordonnanceur de Grille Adaptatif. In: Conférence sur les Architecture Logicielles, Montréal, Canada. Revue des Nouvelles Technologies de l'Information RNTI-L-2, pp. 19–33. Cépaduès-Éditions (2008)

40. Foster, I.: Globus toolkit version 4: Software for service-oriented systems. Journal of Computer Science and Technology 21(4), 513–520 (2006)

41. Erwin, D., Snelling, D.: UNICORE: A Grid computing environment. In: Euro-Par 2001 Parallel Processing, pp. 825–834 (2001)

Data Extraction from Online Social Networks Using Application Programming Interface in a Multi Agent System Approach

Ruqayya Abdulrahman[1,2], Daniel Neagu[1], D.R.W. Holton[1], Mick Ridley[1], and Yang Lan[1]

[1] Department of Computing, University of Bradford, Bradford, U.K
[2] Department of Computer Science, Taibah University, Medinah, Saudi Arabia
`rshabdul@bradford.ac.uk`

Abstract. In recent years, Online Social Networks (OSNs) have attracted a significant increased number of users. New methods for extracting data are required to deal with the real time changes of a huge amount of personal information in OSNs. In the past, we implemented a parser as centralized system to retrieve information from OSN profiles source web pages. One of the drawbacks was that the parser had to be updated to reflect the changes in the profiles' structure. In this paper, we extend our previous work that proposed Online Social Network Retrieval System (OSNRS) to decentralize the retrieving information process from OSN. The novelty of OSNRS, which is based on a Multi Agent System (MAS), is its ability to monitor profiles continuously. The new addition involves replacing the parser with the Application Programming Interface (API) tool to enable OSNRS to be integrated with services that are supported by OSN providers in the absence of the profiles source web page. Also, new algorithms alongside case studies are presented to improve OSNRS. The experimental work shows that using API and MAS simplifies and speeds up tracking the history of OSN profiles. Moreover, combining them with text mining helps us further to understand the dynamic behaviour of OSNs users.

Keywords: data extraction, online social network, agent, multi agent system, formal specification, application programming interface.

1 Introduction

Given the continuously changing information in different web resources including websites and online databases, more research is required to find new methods of information extraction in order to satisfy the users' or applications' requirements. Online Social Networks (OSNs) such as Facebook[1], Myspace[2] and Twitter[3] are

---

[1] `http://www.facebook.com`
[2] `http://www.myspace.com`
[3] `http://www.twitter.com`

N.T. Nguyen (Ed.): Transactions on CCI XI, LNCS 8065, pp. 88–118, 2013.

the top web resources showing a significant increase in changes of personal information on a daily basis [3]. Until now, a small number of researchers have explained their methodology in extracting data from OSN profiles. The majority of researchers either extracted data manually or did not mention how the data has been extracted automatically.

In the past, we have developed a parser for the automated extraction of personal information of OSN profiles and their list of friends, either a list of top friends [5] or a list of all friends [1] based on Breadth First Search algorithm. However, that approach has several sources of limitations which need to be addressed regarding using a centralized system and a parser to retrieve information from the profiles' source pages. Moreover, the previous algorithm did not address monitoring profiles over time.

Thus, this paper aims to continue our previous works [3,2] which developed the Online Social Network Retrieval System (OSNRS) to use Multi Agent System (MAS) technology in order to decentralize the retrieving system. The novelty of OSNRS is in the ability of its MAS to provide real time monitoring of OSN profiles. The new addition involves proposing new algorithms to extract information from OSN rather than relying on the parser that requires a continuous updating to get along with any changes in the structure of the profiles' web page. Most current web service providers, including OSN developers, offer what is called an Application Programming Interface (API) to allow software applications to communicate with each other accurately and securely over the Internet. Using this facility allows OSNRS agents to get the required fields despite the modifications in the representation of the profiles' source web pages. Note that the terms OSN profile, URL address and identity are used interchangeably in this paper.

The structure of rest of the paper is as follows: Section 2 introduces a brief overview of OSN and API. Also it presents different approaches to extract information from OSN and how API could be applied to it. Section 3 explores our methodology to extract information from OSN, which includes the description of the enhanced OSNRS and formalizing the system using Object-Z. Section 4 explains the implementation of OSNRS to extract information from Facebook, currently the most popular OSN, supported with different algorithms to improve OSNRS. The findings and results are stated in Section 5. Finally, Section 6 presents the conclusion and future works.

2 Background and Related Work

Different approaches have been developed in previous years to extract data from OSN. As shown in Figure 1, our proposed system (OSNRS) addresses the following challenges:

1. What: the process of extracting data.
2. Where from: OSN, one of the most recent, popular and dynamic web resources.
3. How: using API and MAS techniques.

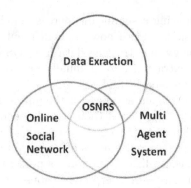

Fig. 1. Venn Diagram of Relevant Fields for OSNRS

In this section, we review the following key literature in means of; web information extraction, OSN, API and various approaches of extracting information from OSN.

2.1   Web Information Representation and Extraction

One of the major concerns of information extraction is how data could be extracted from different web resources such as websites, online databases and services. Before being able to develop tools and methods to extract information, researchers should be aware of how information is represented on the web. The importance of information representation arises when it contributes towards the parsers of search engines in order to interact with the web pages' contents more efficiently. Information representation correlates to how data is available on the web.

Although new approaches to web services for data delivery have been developed via Application Programming Interface (APIs) and Representational State Transfer (REST), the majority of data is still in HTML format [7]. The issue with HTML is that it is designed for semi-structured data which contains information in several forms e.g. text, image, video and audio but their contents may be ill-formed and broken. XML and XHTML impose stricter rules over the structure [17,19].

A useful technique called wrappers is used by [23,26,30]. Wrappers are responsible of converting HTML documents into semantically meaningful XML files. However, wrappers are not efficient though because the programmers have to find the reference point and the absolute tag path of the targeted data content manually. This requires one wrapper for each web site since different sites follow different templates. The effects are increased time consumption and effort from the programmer.

2.2   Online Social Networks

Nowadays, OSN has been used widely by users of various ages. The increased simplicity in accessing the WWW via wireless devices such as laptops and smart phones helps end users to participate in OSNs. This is because it helps them to make new friendships, share their interests even with unknown people, upload photos and distribute their personal information. Thus Lenhart and Madden defined OSN as *"online place where a user can create a profile and build a personal network that connects him or her to other users"* [25].

OSNs have started to be used publicly for communication purpose through Friendster in 2003, although the concept was known since 1960s. Within a few months, Friendster attracted more than 5 million users as reported by [27]. The success of Friendster inspired the creation of other OSNs. By referring to the figure that is drawn by [10], a brief timeline of the launch dates of major OSNs sites was established. The timeline was started by SixDegrees.com in 1997 and progressed with Facebook, Twitter and Windows Live Spaces in 2006.

The growth of OSNs comes along with the expansion of e-commerce. As a result, OSN sites become a major target for companies to advertise their products. Regarding to The Nielsen Company [29], Facebook, Amazon and eBay occupied the top 10 parent companies in United States for Jan 2010 regarding the time users spent on their websites. One of the communication and interaction factors between those users relies on sharing knowledge, interests and experience about products or services. This inspires many studies such as [32,14,13,33,34] to find out the influence of viral marketing and advertising through OSNs in increasing the companies profits rather than the use of traditional marketing.

Though, as any new technology, the rapid expansion of OSNs contributes to raising worries for parents as well as researchers from various disciplines such as psychology, sociology and governments. This is because the youth represents the heart of OSN sites activities and there is little control or protection for such online interaction [34,10]

2.3 Application Programming Interface

IBM has defined an application programming interface (API) as *"a functional interface supplied by the operating system or a separately orderable licensed program that allows an application program written in a high-level language to use specific data or functions of the operating system or the licensed program"* [24]. In contrast, Orenstein simplifies the definition of API to *"a description of the way one piece of software asks another program to perform a service. The service could be granting access to data or performing a specified function"*, i.e. API is a software-to-software interface.

API allows the third-party software developers to access any web services provided by developers of OSN as open source software does. Meanwhile, the privacy of the application's source code is protected as a closed application does. The power of API is in its ability to integrate and interoperate different applications and tools with each other regardless of which programming language is used to write the application or how it was designed.

In this paper, we focused on using API programmatically to enable software developers to extract data from OSN profiles. To the best of our knowledge, few researches reported on using API for extracting data from OSN. For example, [28] collected a data set that include over 11.3 million users and 328 million links from Flickr[4], YouTube[5], LiveJournal[6] and Orkut[7] through integrating their crawlers with the API that is supported by OSN providers. The retrieved data is used to study and analyze the structure of those OSN. The results show that OSN have a much higher fraction of symmetric links as well as higher levels of local clustering. The researchers stated how these properties could affect designing OSN algorithms and applications.

Another research is done by [6] who used API to look for the term university in the name or description of Flickrs groups. The aim was set to investigate if the Flickr community could benefit from the tags that are used within the university image groups. They retrieved a sample of 250 random images uploaded by those groups compound with images' tags. They found out that the images' uploaders pay attention to assign multiple tags (at least four) on each image which is useful for image retrieval purposes.

Mostly when using API, OSN developers return the results in a common Extensible Markup Language (XML) file format or Java Script Object Notation (JSON) objects. Table 1 concludes some of the similarities and differences between them.

2.4   Online Social Networks Information Extraction Approaches

The changes in the users' profiles affect their networks' behaviours and actions and lead to alterations in the pattern analysis. Thus, more attention is required on how information will be collected from OSN websites to deal with the problems of rapid changes [36].

Most of previous research extracted OSN information using non automated approaches such as questionnaires and interviews [4,16,31,18]. However, these approaches are inconvenient nowadays especially with the rapid changes in the size of uploaded information to OSNs on daily basis. Recent research relied more on the automated approach such as crawlers that are used by [35] to extract the interaction between more than 60,000 Facebook users and study the evolution of the activity of a given user profile and friends of their friends over 2 years. They found that the majority interactions from over than 800,000 logged in interactions are generated by a minority of user pairs.

Also [8] used crawlers to extract information from different types of OSN profiles using 3 algorithms in order to study the privacy issues of OSN users; Public listing: to crawl public profiles, false profiles: to crawl searchable profiles via creating false profile, and profile compromise and phishing: to crawl random or specific accounts through malicious applications and phishing attacks.

---

[4] http://www.flickr.com

[5] http://www.youtube.com

[6] http://www.livejournal.com

[7] http://www.orkut.com

Table 1. XML vs JSON

	XML	JSON
Stands for	Extensible Markup Languge	JavaScript Object Notation
Extended from	SGML(Standard Generalized Markup Language)	JavaScript
Developed on	1996	2011
Developed by	World Wide Web Consortium	Douglas Crockford
Official website	http://www.w3c.org/TR/rec-xml	http://json.org
Speed		✓
Simplicity		✓
Extensibility		✓
Interoperability	✓	✓
Openness		✓
Human readable/writeable	✓	
Machine readable/writeable		✓
Resource(CPU/Memory) utilization	✓	
Provide Structure to data	✓	✓
Ease of creating data on server side	✓	
Processing easily form client side		✓
Self description data	✓	✓
Data exchange format		✓
Document exchange format	✓	
Mapping to Object-Oriented program		✓
Internationalization	✓	✓
Adopted by industry	✓	
Ease of debugging and troubleshooting on server side	✓	
Ease of debugging and troubleshooting on client side		✓
Lack of Security	✓	✓

2.5   Online Social Networks Information Extraction Using Agents

To the best of our knowledge, the research in extracting data from OSNs using MAS technology is too little. Chau et. al. in [11] extends the parallel crawler that is designed by Cho et. al in [12]. While Cho works on the static assignment architecture, where each crawler is assigned to a part of the web (in general) to retrieve information and coordinate with each other without a central coordinator, Chau works on dynamic assignment architecture to crawl ebay network using a central coordinator (Master Agent) to control all crawlers.

Chau built a queue to list the ebay pending users who are seen but not visited yet using Breadth First Fashion. The MasterAgent has to ensure that no redundant occurs in visiting the ebay users when each crawler agent sent a request to get the next user from the queue. The crawler agents use multiple threads for crawling then return the extracted information to the Master Agent.

The work presented in this paper is similar to [11]. However, the MasterAgent has to assign which agent has to crawl the next user in the queue of OSN. Moreover, our crawler agents are designed to monitor the profiles of OSN users to build a history for each user rather than visiting each user just once.

3   Overview of the Enhanced System (OSNRS)

In order to appropriately address the problem of information retrieval from OSN, we present below an overview of our contributions to Online Social Network Retrieval System (OSNRS) through illustrating its organizational model and its components, and how these components interact with each other. The formal specification language (Object-Z) is used to give a detailed system description at the abstract level. In addition to this review of current work, we propose new algorithms to extract information from OSN using API which improve our previous approach that is presented in [3].

3.1   Organizational and Structural Models of OSNRS

The Organizational model of the OSNRS is built based on the structure of MAS in general and the selected tool in particular. JADE [8] is selected to implement the OSNRS. In JADE, each agent should live in a running instance of runtime environment called a container. A composite of containers comprises the platform. Although many agents may live in one container, and several containers may compose several platforms, it was decided to simplify the OSNRS environment by:

- Assigning one platform to compose all containers.
- Creating each agent in a unique container.

Figure 2 shows the organizational model of OSNRS and its environment. Note that the number of containers differs in the workstations. Also, the local repository (Database) can be accessed only by the mAg.

The enhanced system, OSNRS could be defined as *"a MAS that consists a finite set of agents (grabber agents) controlled by a special agent called MasterAgent in order to achieve a goal of retrieving and monitoring OSN profiles' information starting from a given OSN profile (seed profile)"*.

Through this paper, the MasterAgent (which is described by mAg) organizes the extraction process, controls agents and saves the retrieved information in the local repository. In contrast, the grabber agent (described by gAg) is used to refer to the agent that is responsible for extracting data from OSN profiles and detecting the updates in these profiles. A kind of gAg called groupManagerAgent

---

[8] http://jade.tilab.com

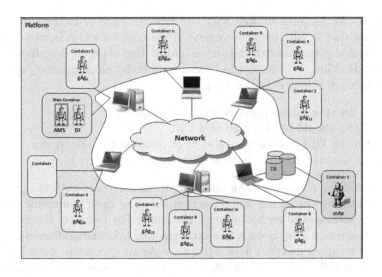

Fig. 2. Organizational Model Of OSNRS

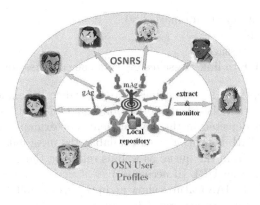

Fig. 3. Structural Model Of OSNRS

(described by gmAg) is added to play a middle role between mAg and gAg as will be detailed in Section 4.3.

Figure 3 shows the structural model of OSNRS. The novelty of the OSNRS is in keeping the gAg monitoring the assigned profile for any updates. Such updates are detected by comparing the current retrieved information with the saved file from a previous extraction. Once a change is captured, the gAg will inform the mAg and send a copy of the updated file.

The master/slave technique is the simplest technique which JADE agents can use for coordination. The mAg (and gmAg if applied as will be described in Section 4.3.3) control and gather all information from the gAgs. They assign tasks to gAgs in order to ensure global coherence of the whole process of extraction.

Several MAS features fit well with the OSNRS. The proactivity feature of mAg means that it initiates achievement of the goal for which it was designed, which as stated is to collect the historical information of the seed profile and that of its friends. Thus, it will start to behave as a stand alone process (autonomy feature) through taking decisions such as:

- how to find the existing gAg in the platforms.
- which gAg should be allocated to a URL.
- whether the URL should be added to or removed from the queue.
- when it should terminate to stop the application.

The sociability feature of mAg allows it to communicate with the gAgs, the user, and other software e.g. database systems. The mAg communicates with gAgs using messages to exchange plans and goals. For instance, the mAg must send a message to a gAg to ask it to stop extracting information from a profile or change the targeted profile. JADE agents use Agent Communication Language (ACL) which follows FIPA ACL standard; the most used and studied agent communication language for messaging format. The most common message attributes include:

- The sender (initialized automatically).
- List of receiver(s).
- Content of the message.
- Conversation_ID: to link messages in the same conversation.

The mAg perceptivity feature comes into play when it balances between the directed goal (to extract historical information from OSN profiles) and the timely response to factors detected in the environment. For example, if a gAg reports that the allocated URL is broken or un available, the mAg must take a decision to remove the URL from the queue as well as releasing the gAg to make it free for allocation to another URL (if necessary).

On the other hand, MAS features show up in the gAg as follows; the autonomy of a gAg allows it to operate as a standalone process to achieve its goal (retrieving and monitoring the profile information) without the direct intervention of users. Also, the autonomy feature of gAg (including gmAg) allows it to decide on the suitable period of time to reactivate itself depending on how active the profile is. This could be done when the gAg compares the retrieved file each time it has been reactivated. If the profile is active, this means that the difference is significant. Accordingly, the gAg will minimize the period of sleeping before reactivating itself and vice versa. In addition, the autonomy of gmAg appears when it has to decide when it has to send the updated files to mAg, how it will control its gAgs and when it has to share knowledges with other gmAgs.

The perceptivity of the gAg permits it to detect changes in profiles and to make decisions as to when it must report the results to the mAg. The mobility feature permits the gAg to be distributed between different machines in the network. The parallelism of gAg helps to speed up the operation of extraction. The scalability of the MAS will facilitate the process of adding new gAg or other

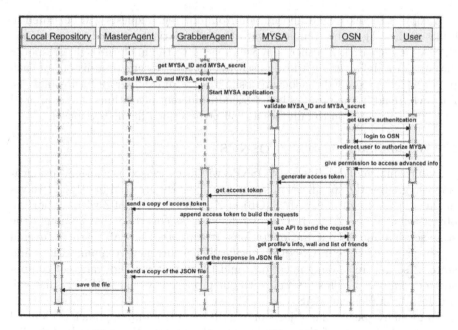

Fig. 4. Sequence Diagram of OSNRS using API

agents for different purposes as required without having to change the existing system. The sociability allows gAg, mAg and users of OSNRS to communicate with each other and exchange their knowledge as explained above.

3.2   The Flow of OSNRS Components

To start the extraction process using API, we create a desktop application called MYSA on the OSN platform. The successful creation of the application generates two parameters; application ID (or API key) and application secret.

As illustrated in Figure 4, when OSNRS starts, the mAg assings a gAg to the seed profile and sends a message which contains the MYSA's parameters to that gAg in order to start the extraction process. The gAg uses these two parameters to establish the connection with the OSN server which will authenticate the user through validating the login cookies which are stored if the user already logged in, or will prompt the user to login if he was not.

Once the user has been authenticated successfully, the user will be redirected to authorize the application to ensure that he knows exactly what type of data and capabilities he authorized MYSA to access. MYSA asks the user for extra permissions in addition to the basic information that OSN provides by default or specified as a public information by the user. When the user authorize MYSA, the application authentication step is approved and the OSN server will generate a response. The response contains the access token accompanied by expires parameter in seconds to be used in every API request.

A copy of the extracted information from different requests will be stored in a file by the gAg to be compared with the information retrieved the next time the gAg is activated. Another copy will be sent to the mAg through a response message. The information will be used by the mAg to create a history of each profile which is recorded in a local repository to be mined later for future analysis. For each friend in the profile's list of friends, different algorithms will be applied to monitor updates on the profiles as will be described in the Sections 4.3.

3.3   Formal Specification of OSNRS

The above informal description of OSNRS will be expressed as a formal specification using Object-Z specification language [15] as Hilaire et.al in [21] and Hayes in [20] selected to formalize MAS. Object-Z is an extension of Z with supporting object-oriented specification. Object-Z has a class construct that combines the state schema and its relevant operations. It supports all MAS features covered by our proposed system (OSNRS) specifically concurrency, communications and state. Also multiple communication could be applied through range of composition operators [9]. This fits well with our proposed system (OSNRS) that will use Java to implement MAS.

To formally specify the objects of OSNRS, a bottom-up approach is required to reflect the nature of these objects and their interaction precisely. Initially, the basic types of the OSNRS and the relations between these types will be identified. Then, the Object-Z Classes of OSNRS will be detailed independently.

3.3.1   OSNRS Basic Types

From the description of OSNRS, the basic types of the enhanced system are defined as:

[Agent, Record, Identity]

where

[Agent] the set of all created agents.
[Record] the set of files that contains the retrieved information from the users' profiles. Time could be included in this type to handle the time of retrieving, updating and saving information.
[Identity] the set of the unique identities of OSN profiles.

As can be seen in Figure 5, the main relations between the basic types of the system are:

1. *visiting*: is a partial injective function from *Agent* to *Identity* to record which identity a gAg is visiting. *visiting* is a partial function, to exclude the initial state and the cases where the identities are retrieved but not assigned to any gAg. It is injective (one-to-one) function because each identity is visited by at most one gAg.
2. *file*: is also a partial injective function from *Identity* to *Record* to represent one record of the identity's extracted data at a time.
3. *history*: records a history of each identity that has been observed over time.

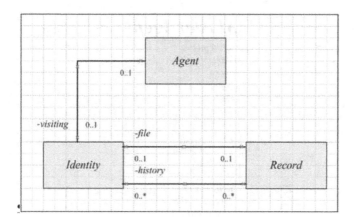

Fig. 5. Relationships between the Basic Types of OSNRS

3.3.2   The Object-Z Classes of OSNRS

Although many compound classes are used to construct OSNRS such as those illustrated in Figure 4, we will focus on the two most important classes: MasterAgent and GrabberAgent. The syntax of these Object-Z classes may contain a constant definition, a state schema, an initial state and finally set of operation schemas.

A. MasterAgent Class

MasterAgent class is the main class in OSNRS that is controlling the extraction process as well as saving a history of extracted profiles' information in a local repository as mentioned earlier. The UML-like class diagram which declares the attributes and the operations of the class can be seen in Figure 6.

The following sub sections describe in detail each component of the class.

3.3.3   State Schema of MasterAgent

The state schema in the class MasterAgent contains four state variables as illustrated below:

1. *known*: is an injective sequence of all identities that are known to the MasterAgent, either the identity has been retrieved as a friend in the friends list or entered by the user. Each identity in *known* appears once.
2. *waitingResponse*: is a set of identities that gAgs have been assigned to, but whose details have not yet been retrieved.
3. *visiting*: is a partial injective function from *Agent* to *Identity* as described earlier in Section 3.3.1.
4. *current* and *next*: are tuples. *current* is used to determine the sequence identities that are being retrieved in this iteration level while *next* represents the identities that will be retrieved in the next iteration level. In other words, when the extraction process moves from an iteration level to another level, the contents of *next* will become as *current*; then *next* will contain the list of friends of identities in the *current*.

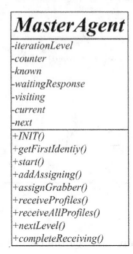

Fig. 6. MasterAgent UML-like Class Diagram

The state invariants points out two roles of *known*; first of all, it has to record all retrieved identities in addition to the initial identity provided by the user, denoted (*seed?*). The second role is to filter the retrieved identities before adding them to the existing list. Thus, all identities in *next*, *current* and *waitingResponse* are taken from *known*.

Since in any level, the identities that are in *next* will become in the *current* for the next levels, the final invariant will ensure that the identity will not be extracted twice by different gAgs. When an object of the class MasterAgent is initialized, no identity has been added to the sequence *known* yet. Consequently, no identities are waiting for response in *waitingResponse*. The *visiting* variable is empty set because the gAg has not been allocated to any identity. Therefore, *current* and *next* are tuples that, on level -1, contain empty sequences of identities that are being or will be retrieved respectively.

3.3.4   MasterAgent Operations
The following seven main operations are specifying the transitions the Master-Agent class can undergo. They are as below:

I. The *start* Operation

When an object of the MasterAgent is established for first time, the *start* operation will initialize all state variables as described in *INIT* schema, then it will be composed sequentially with the *getFirstIdentity* operation as will be explained below.

II. The *getFirstIdentity* Operation

When the user entered the identity *seed?*, the mAg has to add this *seed?* to the sequence *known* that holds all identities that the system will retrieve their information. Also, *seed?* will be added to *waitingResponse* set to indicate that this identity is assigned to a gAg but has not retrieved its information yet.

Consequently, the tuple (*seed?*, 0) is allocated to *next* to specify that *seed?* profile will be extracted in the next iteration.

III. The *assignGrabber* Operation

As described in Section 3.2, the first operation the MasterAgent object has to do when it receives the *seed?* from the user is to assign it to a gAg. The *assignGrabber* is a recursive operation that, when composed with the *addAssigning* operation (as detailed below), will continue until all identities in the sequence *next* will be allocated to gAgs.

IV. The *addAssigning* Operation

To express the *addAssigning* operation, we have to assent that there exists a gAg from the class GrabberAgent which has not been assigned before to any identity. This gAg will be allocated to the head of the sequence of identities in the *next* (denoted *id!*).

Consequentially, the *id!* will be removed from the sequence and the rest elements will be shifted to get ready for allocation to the subsequent gAg. Note that the iteration level will not increment in this operation. Once *addAssigning* accomplished successfully, the identity (*id!*) will be passed to GrabberAgent class for retrieving its information as will be explained in Section 3.3.6.

V. The *receiveAllProfiles* Operation

The *receiveAllProfiles* operation that was called in *AssignGrabber* operation is also a recursive composition operation, that is composed with the receiveProfiles operation. But, it will not move to the next level of iteration until information about all identities in *waitingResponse* is returned.

VI. The *receiveProfiles* Operation

When the mAg receives information of a profile (*id?*) from a gAg, it has to pass this information (*newRec?* and *newFriends?*) to the LocalRepository class (see Figure 4) for updating the history of the profile *id?*. It is necessary that the mAg has to filter the *newFriends?* to remove identities that already exist in *known*. Then it has to remove this *id?* from the *waitingResponse*.

VII. The *completeRetrieving* Operation

This operation either applies the *nextLevel* operation recursively or exits and finishes the iteration if the counter of *iterationLevel* matches the number in the second *next* relation.

8. The nextLevel operation

When the mAg has completed retrieving all information about all identities in the *waitingResponse* set, i.e. complete the iteration level, it has to move for the next level of iteration. In this stage, all identities in *next* will be copied to both *waitingResponse* and *current* to start new level of retrieving information. The complete MasterAgent object is displayed below in Figure 7.

MasterAgent

$iterationLevel : \mathbb{N}$ $counter : \mathbb{N}$

$known$: seq $Identity$
$waitingResopnse$: $\mathbb{P}$ $Identity$
$visiting$: $Agent \rightarrowtail Identity$
$current, next$: (seq $Identity \times \mathbb{Z}$)

$waitingResponse \subseteq \mathrm{ran}\, known$
$\mathrm{ran}\, firstnext \subseteq \mathrm{ran}\, known$
$\mathrm{ran}\, firstcurrent \subseteq \mathrm{ran}\, known$
$\mathrm{disjoint}\langle \mathrm{ran}\, firstcurrent, \mathrm{ran}\, firstnext \rangle$

INIT

$known = \langle \rangle$
$waitingResponse = \varnothing$
$visiting = \varnothing$
$current = (\langle \rangle, -1)$
$next = (\langle \rangle, -1)$

$start \;\widehat{=}\; \text{INIT} \;\mathbin{\raise2pt{\tiny\S}}\; getFirstIdentity$

getFirstIdentity
$\Delta known, waitingResponse, next$
$seed?$: $Identity$

$seed? \notin \mathrm{ran}\, known$
$known' = \langle seed? \rangle \;\wedge$
$waitingResponse' = \{seed?\}$
$next' = (\langle seed? \rangle, 0)$

$assignGrabber \;\widehat{=}\; addAssigning \;\mathbin{\raise2pt{\tiny\S}}\; assignGrabber$
$[]$
$[firstnext = \langle \rangle]\, receiveAllProfiles$

addAssigning
$\Delta visiting, next$
$id!$: $Identity$

$\exists\, gAg : GrabberAgent \bullet gAg \notin \mathrm{ran}\, visiting \Rightarrow$
$visiting' = visiting \cup gAg \mapsto id! \;\wedge$
$next' = (tail\, firstnext, secondnext) \;\wedge$
$id! = head(firstnext) \;\wedge$
$gAg.receiveID$

$receiveAllProfiles \;\widehat{=}\; receiveProfiles \;\mathbin{\raise2pt{\tiny\S}}\; receiveAllProfiles$
$[]$
$[waitingResponse = \langle \rangle]\, completeRetrieving$

Fig. 7. Formal Description of MasterAgent Class

$MasterAgent(cont)$

$receiveProfiles$

$\Delta known, waitingResponse, next$
$id? : Identity$
$newRec? : Record$
$newFriends? : seq\ Identity$

$waitingResponse \neq \varnothing$
$\textbf{let}\ filtered : seq\ Identity\ \wedge$
$filtered = (newFriends? \upharpoonright (Identity \setminus ran\ known))$
$known' = known \frown filtered$
$next' = newFriends?$
$waitingResponse' = waitingResponse \setminus \{id?\}$

$completeRetrieving \;\widehat{=}\; nextLevel \,\fatsemi\, assignGrabber$
$[]$
$[secondnext = iterationLevel \vee \#known \geq counter]INIT$

$receiveProfiles$

$\Delta known, waitingResponse, next$
$id? : Identity$
$newRec? : Record$
$newFriends? : seq\ Identity$

$waitingResponse \neq \varnothing$
$\textbf{let}\ filtered : seq\ Identity\ \wedge$
$filtered = (newFriends? \upharpoonright (Identity \setminus ran\ known))$
$known' = known \frown filtered$
$next' = newFriends?$
$waitingResponse' = waitingResponse \setminus \{id?\}$

$nextLevel$

$\Delta current, next, waitingResponse$
$id! : seq\ Identity$

$secondnext \neq iterationLevel$
$counter \leq \#known$
$next' = (\langle\,\rangle, secondnext + 1)\ \wedge$
$current' = next\ \wedge$
$waitingResponse' = ran\ firstnext\ \wedge$
$mAg.assignGrabber$

Fig. 7. *(continued)*

3.3.5   GrabberAgent Class

The second main class in OSNRS is the GrabberAgent. The objects of Grabber-Agent class in OSNRS could be considered as the worker bees in the bees' colony. This is because the GrabberAgent objects have to travel through network, visit the OSN profile, extract data, monitor updates and send regular reports to Mas-terAgent class. The Object-Z specification of the class GrabberAgent is given in Figure 8.

The state schema for the GrabberAgent class contains three state variables:

1. *myID*: denotes the unique profile identity that the gAg is assigned to.
2. *myRec*: denotes the personal information of the profile identity.
3. *myFriends*: denotes the list of friends of the profile identity.

The predicate of the state schema is true if and only if each identity is not a friend of itself.

When an object of the class GrabberAgent is created, it would not be allo-cated to any identity yet. Accordingly, neither the personal information of the profile identity, nor its list of friends is known. Thus, the state variable *myRec* is initialized to empty set while *myID* and *myFriends* are set to empty sequences.

3.3.6   GrabberAgent Operations

Although GrabberAgent has only three main operations, they are what OSNRS based upon. They are as following:

I. The *receiveID* Operation

As mentioned earlier in Section 3.3.4, when the mAg assigns a gAg to an iden-tity, The GrabberAgent class will get this identity through *receiveID* operation. The *receiveID* has to compose several operations sequentially.

II. The *updateID* Operation

The *upadateID* operation is called to replace the old identity that the gAg is allocated to with a new identity (*id?*). This is important to re-use the gAg in cases where the identity that it is assigned to was removed from the sub network of OSN, or a gAg is assigned to a new profile.

III. The *updateInfo* Operation

The *updateInfo* operation is involved periodically after the gAg has been assigned an *id?*. Since the retrieved information is coming from an external database (OSN server), we will assume that there exists a record (*newRec?*) that contains all personal information about *myID* and a sequence of identities (*newFriends?*) that holds a list of *myID*'s friends.

Either the *newRec?* or *newFriends?*, or both should be distinguished from the existing *myRec* and *myFriends* respectively in order to be replaced. As a result, the new information in *myRec*, *myFriends* accompanied with *myID* will be sent back to the MasterAgent class through *rec!*, *friends!* and *ID!* consecutively when mAg.*receiveProfile* operation is involved.

```
┌─ GrabberAgent ──────────────────────────────────────────────────────
│
│  ┌──────────────────────────┐  ┌─ INIT ──────────────────────────┐
│  │ myID : Identity          │  │ myID = ⟨ ⟩                       │
│  │ myRec : Record           │  │ myRec = ∅                        │
│  │ myFriends : seq Identity │  │ myFriends = ⟨ ⟩                  │
│  │ ──────────────────────── │  └─────────────────────────────────┘
│  │ myID ∉ ran myFriends     │
│  └──────────────────────────┘
│
│  ┌─ updateInfo ──────────────────────────────────────────────────┐
│  │ rec! : Record                                                  │
│  │ friends! : seq Identity                                        │
│  │ id! : Identity                                                 │
│  │ ────────────────────────────────────────────────────────────  │
│  │ ∃ newRec? : Record, newFriends? : seq Identity •               │
│  │ newRec? ≠ myRec ∨ newFriends? ≠ myFriends ⇒                    │
│  │ myRec‘ = newRec? ∧                                             │
│  │ myFriends‘ = newFriends? ∧                                     │
│  │ rec! = myRec‘ ∧                                                │
│  │ friends! = myFriends‘ ∧                                        │
│  │ id! = myID ∧                                                   │
│  │ mAg.receiveProfile                                             │
│  └────────────────────────────────────────────────────────────────┘
│
│  receiveID ≙ INIT ⨟ updateID ⨟ updateInfo
│
│  ┌─ updateID ────────────────────────────────────────────────────┐
│  │ ΔmyID                                                          │
│  │ id? : Identity                                                │
│  │ ──────────────────────────────────────────────────────────── │
│  │ myID‘ = id?                                                   │
│  └────────────────────────────────────────────────────────────────┘
│
└─────────────────────────────────────────────────────────────────────
```

Fig. 8. Formal Description of GrabberAgent Class

4   Implementation of OSNRS

As stated in the introduction, the novelty of OSNRS is in its ability to monitor the real time changes in the OSN profiles. The previous OSNRS algorithm [3] was relying on a parser to extract information from OSN web pages' source and monitor the updates. Using such a parser is inconvenient or impossible in some cases where the OSN developers (e.g. Facebook) do not provide the source web pages of OSN profile. Instead, they allow to access information via an API. Thus, the following sections describe how to implement the improved OSNRS through using API.

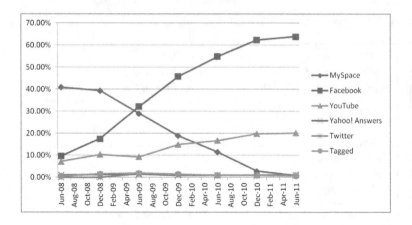

Fig. 9. Top Online Social Network (source:Hitwise)

4.1   Setting up the OSNRS Environment

OSNRS is implemented using Java Agent DEvelopment framework (JADE). According to JADE, agents should be created in container(s) and a set of active containers comprise a platform. Although JADE allows several agents to live in one container as well as several containers could forming many platforms as shown in Figure 2, to simplify our system we create each agent of OSNRS in a container. The set of containers are distributed on two connected workstations to compose one platform.

4.2   Choosing the OSNRS Sample

In our previous works, we created a parser to crawl the web pages of Mysapce profiles. However, Myspace has lost market share dramatically through last three years. In contrast, Facebook has increased strongly as reported by [22] based on the United States market share of visits (see Figure 9). Therefore, Facebook was selected as the domain for OSNRS data extraction.

Facebook puts the users' privacy at their top priority. Thus, third-parties are not allowed to access users' profiles but through API. To use Facebook Graph API, we created a Facebook application called MYSA as a desktop application rather than a web application to be compatible with our existing OSNRS. Facebook demands its application (in our case MYSA) to be authorized and authenticated by their users to ensure that the users give their permission to MYSA to access their information.

Although the experiment has been tested on the Facebook real accounts of the authors and their networks, a mock network consisted of more than 20 Facebook users has been used as OSNRS's sample to avoid privacy and ethical issues. The mock network is taken from a project called "The Artemis" developed at the University of Bradford. Figure 10 shows a screen shot of some of the mock

Fig. 10. A Sample of Mock Network Profiles

network profiles. The users of the mock network are connected partially with each other in a random relationship.

Figure 11 shows the mock network where the vertices represent the users or profiles and the edges are the relationships. Note that just over half of the sample (55%) are females, symbolized by rectangle vertices while males are symbolized by a circle vertices. 41% of users are stated as classmates while 22% are connected by a family relationship such as fatherhood, motherhood and brotherhood (represented by bold edges). The rest are friends or having a share interests.

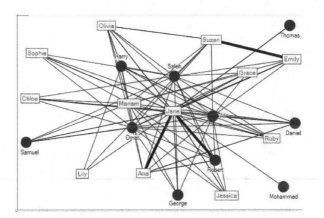

Fig. 11. Mock Network for OSNRS

4.3   Setting and Implementing OSNRS Algorithms

Following setting up JADE environment and building MYSA application, this section proposes algorithms to improve the previous OSNRS presented in [3]

```
Given:
-    Parameters of OSN Desktop application(MYSA_ID and MYSA_secrect).
-    Created agent, MasterAgent (mAg).
-    Ag set of grabberAgent(s).
-    u: user of OSN.

Input:
-    u enters his userName and password (seed profile).
-    u gives a permission to authorize MYSA.

Output:
-    Profile's historical information such as personal information,
     home, wall and list of friends (as permitted).

Steps:
The mAg will:
-    starts MYSA application.
-    Call OSN_authentication(MYSA_ID, MYSA_secrect).
-    Get the access_token from MYSA.
-    if(gAg ∈ Ag && gAg not allocated to ID)
       - Allocate gAg to the logged in user (ID).
-    Send the access_token to the gAg.
-    Append access_token to build the API queries.
-    Extract the profile's information of u.
-    Save a copy of the information in file.
-    Send a copy of the file to the MasterAgent.
-    Go to one of the sub-algorithms.


OSN_authentication(MYSA_ID, MYSA_secrect)
Begin
    The OSN has to validate the MYSA_ID and MYSA_secret
    if (MYSA_ID && MYSA_secret) valid then
    begin
        get authentication of u
        if(u is logged in)then
          check the cookies
        else
          redirect u to login dialog of the OSN
        if (u is authenticated)then
            redirect u to authorize MYSA dialog.
            if (MYSA is authorized)then
                generate access_token.
    End if
    return access_token.
End
```

Fig. 12. General Algorithm of OSNRS with API

by applying API within MAS approach accompanied by case studies. Figure 12
illustrates the pseudocode of the general algorithm of the improved OSNRS.

The algorithm matches the system informal description in Section 3.2, where
the mAg is created to establish the OSNRS and looked for existing gAg to send
the MYSA's parameters: the application ID and secret key. These parameters are

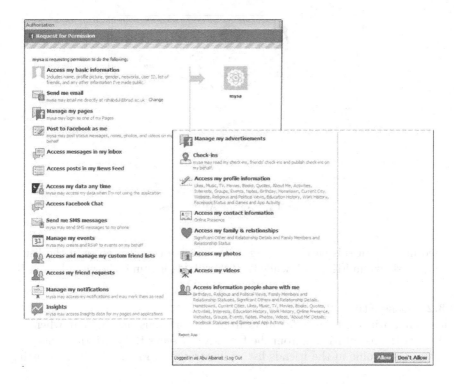

Fig. 13. MYSA's Authorization Request

used to get the users authentication and MYSA's authorization. Authorization is obtained when MYSA asks users for permissions to access extra information beside the basic information that Facebook developers provide by default, and the information that Facebook users set it as public. Figure 13 illustrates the required permissions. The process is continued as described in Section 3.2 until the mAg gets the profile's information and the list of friends from the first gAg accompanied with the access token.

Starting from this point, we developed three different algorithms to deal with how to extract friends information due to some of OSNs (e.g. Facebook) do not allow their applications to access the friends of friends of the logged in user information unless the user and his friends give their permissions. These algorithms will be compared through case studies to find out the best algorithm in terms of the speed of the extraction process, the accuracy of the extracted information and the performance of the agents.

4.3.1   Case Study 1
In the first case study, when the mAg gets the list of profile's friends, it will seek the platform to find another gAg to retrieve all possible information and to monitor the updates on the profiles of all friends in the list of friends of the

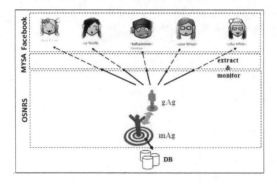

Fig. 14. gAg for All Profiles

logged in user. In this case, one gAg is responsible for all profiles in the list of friends as shown in Figure 14, while the first sub algorithm is shown in Figure 15.

4.3.2   Case Study 2

The scenario in case study 2 differs from the previous one in that when the mAg gets the list of friends from the first gAg, it seeks the platform to assign a gAg to each profile in the friends list as illustrated in Figure 16. If the existing

```
Steps:

The mAg will
-     look for a gAg in P.
-     if ( gAg ∈ P && ((gAg,ID)∉ Assigned ))then
-     begin
-         Assigned = Assigned ∪ (gAg,ID).
-         send access_token to gAg and call
            gAg.grabInfo (access_tokne).
-         get the file from gAg.
-         save a copy of file.
-     end
-     re-activate gAgᵢ at the specified time.

grabInfo(access_token)
      the gAg will
-     begin
-       re-activate the gAg.
-       for each ID in L do
-           append access_token to build the API queries.
-           extract the ID's basic and public information.
-           extract the ID's wall information.
-           save the information in a file.
-           return a copy of the file to mAg.
-           keep listening to the wall.
-           go to sleep for a while.
-     end
```

Fig. 15. OSNRS Sub Algorithm 1

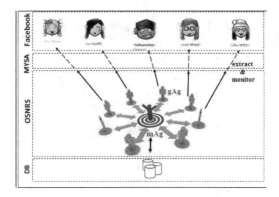

Fig. 16. A gAg for Each Profile

gAgs are not enough or not found in first place, mAg will create gAgs as much as required to match the number of friends list. Consequently, each gAg will act as the gAg in case study 1 and communicate with each other to exchange knowledge. Figure 17 shows the second sub algorithm.

```
Steps:

-      for each ID_i in L do
-      begin
-         look for a gAg in P
-         if( ( gAg_i ∈ P) && ( gAg_i ∉ Assigned ))then
-            begin
-               Assigned = Assigned ∪ (gAg_i,ID_i).
-               send access_token to gAg_i and call
                  gAg.grabInfo (access_tokne).
-               get the file from gAg_i.
-               save a copy of file.
               end
-         re-activate gAg_i at the specified time.
-      end

grabInfo(access_token)
-      begin
-         re-activate the gAg.
-         for each ID in L do
-            append access_token to build the API queries.
-            extract the ID's basic and public information.
-            extract the ID's wall information.
-            save the information in a file.
-            return a copy of the file to mAg.
-            keep listening to the wall.
-            go to sleep for a while.
-      end
```

Fig. 17. OSNRS Sub Algorithm 2

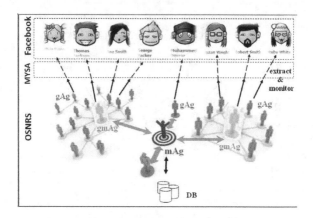

Fig. 18. Two level of Controlling Profiles

4.3.3   Case Study 3

Since Facebook does not give permissions to applications to retrieve friends of friends list, we studied a scenario where two or more users could login to Facebook to allow MYSA access their information. These users should be in the list of friends of the seed user. In this way, we allow MYSA to move deeply in the social network. For this scenario, we have two level of controlling rather than one as we did previously. The mAg will control the whole system but instead of having control with all gAgs, a second level of controlling associated with a special kind of gAg that is called a groupManagerAgent (gmAg). Figure 18 shows the structure of OSNRS when gmAg is included.

The gmAg has to play two roles. Firstly, it will extract all information related to the profile which it is assigned to including the list of friends and return it to the mAg. Secondly, it will get the filtered list of friends (profiles which have not been crawled yet by other gAgs) from the mAg and allocate new gAg to each profile. The gmAg will be the link to coordinate the communication between the mAg and the gAgs.

For the all three case studies when the gAg extracts the information, it will send all information back to the mAg in JSON files in order to build a history of each profile. Each file name combines the Facebook profile unique ID (either name or ID number)+ the file contents such as info, wall, searchPeople, searchPost, etc + the retrieving time in (YYYYMMDDhhmmss) format, where 'Y' used for year, 'M' for month, 'D' for day, 'h' for hour, 'm' for minute and 's' for second. Such a format guarantees that the file is unique and simplifies the saving of a history of each profile.

Although mAg can save the returned files in database tables as we did in [3], we found it is inappropriate and time consuming to save the information that is in JavaScript Object Notation (JSON) files in database tables, then queries the tables to extract the information and save it back to JSON file. Thus, we decided to leave the information in JSON files especially when JSON is considered to be smaller, faster and easier to parse than XML as shown previously in Table 1.

Also, gAg keeps a copy of information to use for comparing the results the next time the gAg is activated. Moreover, gAg would be able to search on profiles' walls for keywords that is required when mining data. For example, the agent could look for keywords such as birthday then compare if the date of birth that is retrieved from the user's information matches the greetings date that friends wrote on the user's wall. The result would be used to help calculate the vulnerability of the user based on how the friends could leak the user's information as explained in [1].

For searching purposes, OSNRS will create a special agent to look for the requested keyword(s). This agent will communicate with all gAgs in the platform. Having a special search agent is more appropriate than letting each gAg to scan all other walls to get the results.

5   Findings and Results

The aim of the experimental work is to improve OSNRS through using API tool. Thus, the table below shows some of differences between retrieving information using the parser in the first version of OSNRS, which is implemented in [3], and in the improved version of OSNRS in this paper through retrieving information using API.

Although generally speaking, we do not have to concern about the structure of data representation when API is used, it was surprising that we still have to update OSNRS code to deal with the changes in the structure of data in JSON file.

Figure 19 shows some of the retrieved information in JSON files. The values of "television" and "friends" attributes have changed in week 2 of parsing to include the sub attribute "paging" in addition to "data". This affects the mAg when it has to parse the returned JSON file to get the list of friends.

The data in Figure 20 represents information about 15 seeds profiles of the mock network. These profiles have been parsed, as described in case studies 1 and 2, in order to calculate the required time for retrieving 2 JSON files for the seeds profiles and their friends.The first file contains all possible information as shown in Figure 13 while the second file contains the profiles' walls.

The average retrieved files is 15 file with average size of 60 kilobyte. Note that used port to transfer files are different from the port of exchanging messages between agents to speed the retrieving process.

Table 2. parser vs API in developing OSNRS

Criterion	Parser	API
Allow accessing Friends of Friends (FoF) automatically	Y	
Allow direct information access		Y
More concern about privacy		Y
Require updating system regarding structure's changes	Y	
Allow accessing basic information (even for private profiles)	Y	Y

```
{
  "id":"100002623662152",
  "name":"Jane Smith",
  "first_name":"Jane",
  "last_name":"Smith",
  "picture":"http://profile.ak.fbcdn.net
  /hprofile-ak-snc4
  /174530_100002623662152_7581634_q.jpg",
  "television": {
     "data": [2]
  },
  "videos": {
     "data": {
     }
  },
  "friends": {
     "data": [16]
  }
}
```
```
{
  "id":"100002623662152",
  "name":"Jane Smith",
  "first_name":"Jane",
  "last_name":"Smith",
  "picture":"http://profile.ak.fbcdn.net
  /hprofile-ak-snc4
  /174530_100002623662152_7581634_q.jpg",
  "television": {
     "data": [2],
     "paging": {...}
  },
  "videos": {...},
  "friends": {
     "data": [16],
     "paging": {...}
  }
}
```

Fig. 19. Parts of JSON Results (Week 1 on The Left, Week 2 on The Right)

From the accompanied charts (Figures 21 and 22), we can notice the significant difference in the time for retrieving the same number and size of files. The most interesting profile is profile number 14 which has the largest number of friends in the mock network. In case study 1, where one gAg is used to retrieve and monitor all profiles, a sharp increase in the required time for extracting information compared with case study 2 where one gAg is responsible about each profile.

Moreover, the most consuming time in case study 2 related to allocating gAgs and other processing issues. This supported by the fact that the average of 10

Seed Profile	No of Friends	Total Files (units)	Total Size (KB)	Total Process (seconds)	
				Case Study 1	Case Study 2
1	7	16	61	66	45
2	5	12	56	51	42
3	6	14	60	58	42
4	5	12	56	52	43
5	5	12	56	51	39
6	5	12	56	53	43
7	5	12	56	53	39
8	8	18	75	82	50
9	6	14	58	62	44
10	6	14	58	59	43
11	7	17	60	71	46
12	7	16	62	64	49
13	5	12	56	53	40
14	16	34	74	119	79
15	6	14	60	59	41

Fig. 20. Some Results of Case Studies 1 and 2

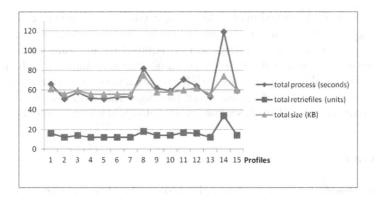

Fig. 21. Case Study 1 Result

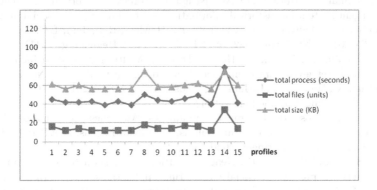

Fig. 22. Case Study 2 Result

files could be retrieved within 3 seconds. In contrast, each file requires around 1 second to be retrieved in case study 1. Therefore, regarding these results, sub algorithm 2 is better than sub algorithm 1.

6   Conclusions and Future Work

This paper continued previous work in developing OSNRS to retrieve information from OSN profiles and monitor the updates in those profiles. We improved OSNRS through using API in order to address the drawback of the parser that had to be updated to reflect the changes in the structure of the profile.

We presented two algorithms aligned with case studies to improve OSNRS. Since the accurately and correctly retrieving and monitoring OSN profiles is not affected by the size of the sample, the initial results of the experimental work is promising especially when using the sub algorithm 2. Thus, a sub algorithm 3 will be implemented to enhance OSNRS.

However, several limitations to this work need to be acknowledged. Firstly, when using API, the retrieving process is limited by the allowance of expire time of the access token. More investigation is required to use some features such as off line permission in order to allow continuous monitoring. Secondly, privacy issues should be addressed to apply OSNRS to a real network with larger number of seeds profiles and friends (some work has been started).Last but not least, further study is required to check the file contents and evaluate agents.

References

1. Abdulrahman, R., Alim, S., Neagu, D., Ridley, M.: Algorithms for data retrieval from online social network graphs. In: 2010 IEEE 10th International Conference on Computer and Information Technology (CIT), pp. 1660–1666. IEEE (2010)
2. Abdulrahman, R., Holton, D.R.W., Neagu, D., Ridley, M.: Formal specification of multi agent system for historical information retrieval from online social networks. In: O'Shea, J., Nguyen, N.T., Crockett, K., Howlett, R.J., Jain, L.C. (eds.) KES-AMSTA 2011. LNCS, vol. 6682, pp. 84–93. Springer, Heidelberg (2011)
3. Abdulrahman, R., Neagu, D., Holton, D.R.W.: Multi agent system for historical information retrieval from online social networks. In: O'Shea, J., Nguyen, N.T., Crockett, K., Howlett, R.J., Jain, L.C. (eds.) KES-AMSTA 2011. LNCS, vol. 6682, pp. 54–63. Springer, Heidelberg (2011)
4. Acquisti, A., Gross, R.: Imagined communities: Awareness, information sharing, and privacy on the facebook. In: Danezis, G., Golle, P. (eds.) PET 2006. LNCS, vol. 4258, pp. 36–58. Springer, Heidelberg (2006)
5. Alim, S., Abdul-Rahman, R., Neagu, D., Ridley, M.: Data retrieval from online social network profiles for social engineering applications. In: International Conference for Internet Technology and Secured Transactions, ICITST 2009, pp. 1–5. IEEE (2009)
6. Angus, E., Thelwall, M., Stuart, D.: General patterns of tag usage among university groups in flickr. Online Information Review 32(1), 89–101 (2008)
7. Baumgartner, R., Campi, A., Gottlob, G., Herzog, M.: Chapter 6: Web data extraction for service creation. In: Ceri, S., Brambilla, M. (eds.) Search Computing. LNCS, vol. 5950, pp. 94–113. Springer, Heidelberg (2010)
8. Bonneau, J., Anderson, J., Danezis, G.: Prying data out of a social network. In: International Conference on Advances in Social Network Analysis and Mining, ASONAM 2009, pp. 249–254. IEEE (2009)
9. Bowen, J.P.: Formal speci cation and documentation using Z: A case study approach, vol. 66. International Thomson Computer Press (1996)
10. Boyd, D.M., Ellison, N.B.: Social network sites: Definition, history, and scholarship. Journal of Computer-Mediated Communication 13(1), 210–230 (2008)
11. Chau, D.H., Pandit, S., Wang, S., Faloutsos, C.: Parallel crawling for online social networks. In: Proceedings of the 16th International Conference on World Wide Web, pp. 1283–1284. ACM (2007)
12. Cho, J., Garcia-Molina, H.: Parallel crawlers. In: Proceedings of the 11th International Conference on World Wide Web, pp. 124–135. ACM (2002)
13. Clemons, E.K., Barnett, S., Appadurai, A.: The future of advertising and the value of social network websites: some preliminary examinations. In: Proceedings of the Ninth International Conference on Electronic Commerce, pp. 267–276. ACM (2007)

14. Domingos, P.: Mining social networks for viral marketing. IEEE Intelligent Systems 20(1), 80–82 (2005)
15. Duke, R., Rose, G.: Formal Object Oriented Speci cation Using Object-Z. Palgrave Macmillan (2000)
16. Dwyer, C., Hiltz, S.R., Passerini, K.: Trust and privacy concern within social networking sites: A comparison of facebook and myspace. In: Proceedings of AMCIS. Citeseer (2007)
17. Gao, Y., Yuan, F., Zhang, M.: Data extraction based on index path in web. In: 2010 Second International Workshop on Education Technology and Computer Science, pp. 157–160. IEEE (2010)
18. Gibson, R.: Who's really in your top 8: network security in the age of social networking. In: Proceedings of the 35th Annual ACM SIGUCCS Fall Conference, pp. 131–134. ACM (2007)
19. Gyarmati, L., Trinh, T.A.: Measuring user behavior in online social networks. IEEE Network 24(5), 26–31 (2010)
20. Hayes, I., Flinn, B.: Specification case studies. Prentice-Hall International (1987)
21. Hilaire, V., Koukam, A., Gruer, P., Müller, J.-P.: Formal specification and prototyping of multi-agent systems. In: Omicini, A., Tolksdorf, R., Zambonelli, F. (eds.) ESAW 2000. LNCS (LNAI), vol. 1972, pp. 114–127. Springer, Heidelberg (2000)
22. Experian Hitwise. Top 10 social networking websites & forums (August 2011), http://www.marketingcharts.com/interactive/ top-10-social-networking-websites-forums-september-2011-19248/
23. Hong, J.L., Siew, E.G., Egerton, S.: Viwer-data extraction for search engine results pages using visual cue and dom tree. In: 2010 International Conference on Information Retrieval & Knowledge Management (CAMP), pp. 167–172. IEEE (2010)
24. IBM. Quickstudy: Application programming interface (api), http://publib.boulder.ibm.com/infocenter/iseries/v7r1m0/ index.jsp?topic=%2Fapiref-%2Fapi.htm
25. Lenhart, A., Pew Internet & American Life Project: Social networking websites and teens: An overview. Pew/Internet (2007)
26. Liu, B., Zhai, Y.: NET – A system for extracting web data from flat and nested data records. In: Ngu, A.H.H., Kitsuregawa, M., Neuhold, E.J., Chung, J.-Y., Sheng, Q.Z. (eds.) WISE 2005. LNCS, vol. 3806, pp. 487–495. Springer, Heidelberg (2005)
27. Mika, P.: Flink: Semantic web technology for the extraction and analysis of social networks. Web Semantics: Science, Services and Agents on the World Wide Web 3(2-3), 211–223 (2005)
28. Mislove, A., Marcon, M., Gummadi, K.P., Druschel, P., Bhattacharjee, B.: Measurement and analysis of online social networks. In: Proceedings of the 7th ACM SIGCOMM Conference on Internet Measurement, pp. 29–42. ACM (2007)
29. NielsenWire. Top 10 social networking websites & forums (November 2010)
30. Palmieri Lage, J., da Silva, A.S., Golgher, P.B., Laender, A.H.F.: Automatic generation of agents for collecting hidden web pages for data extraction. Data & Knowledge Engineering 49(2), 177–196 (2004)
31. Strater, K., Richter, H.: Examining privacy and disclosure in a social networking community. In: Proceedings of the 3rd Symposium on Usable Privacy and Security, pp. 157–158. ACM (2007)
32. Subramani, M.R., Rajagopalan, B.: Knowledge-sharing and infiuence in online social networks via viral marketing. Communications of the ACM 46(12), 300–307 (2003)

33. Swamynathan, G., Wilson, C., Boe, B., Almeroth, K., Zhao, B.Y.: Do social networks improve e-commerce?: a study on social marketplaces. In: Proceedings of the First Workshop on Online Social Networks, pp. 1–6. ACM (2008)
34. Trusov, M., Bucklin, R.E., Pauwels, K.: Effects of word-of-mouth versus traditional marketing: Findings from an internet social networking site. Journal of Marketing 73(5), 90–102 (2009)
35. Viswanath, B., Mislove, A., Cha, M., Gummadi, K.P.: On the evolution of user interaction in facebook. In: Proceedings of the 2nd ACM Workshop on Online Social Networks, pp. 37–42. ACM (2009)
36. Ye, S., Lang, J., Wu, F.: Crawling online social graphs. In: 2010 12th International Asia-Pacific Web Conference (APWEB), pp. 236–242. IEEE (2010)

Cooperatively Searching Objects
Based on Mobile Agents

Takashi Nagata[1], Munehiro Takimoto[1], and Yasushi Kambayashi[2]

[1] Department of Information Sciences, Tokyo University of Science, Japan
[2] Department of Computer and Information Engineering,
Nippon Institute of Technology, Japan

Abstract. This paper presents a framework for controlling multiple robots connected by communication networks. Instead of making multiple robots pursue several tasks simultaneously, the framework makes mobile software agents migrate from one robot to another to perform the tasks. Since mobile software agents can migrate to arbitrary robots by wireless communication networks, they can find the most suitably equipped and/or the most suitably located robots to perform their task. In this paper, we propose a multiple robot control approach based on mobile agents for searching targets as one of the effective examples. Though it is a simple task, it can be extended to any other more practial examples, or be used as an element of a real application because of its simplicity. We have conducted two kinds of experiments in order to demonstrate the effectiveness of our approach. One is an actual system with three real robots, and the other is a simulation system with a larger number of robots. The results of these experiments show that our approach achieves reducing the total time cost consumed by all robots while suppressing the energy consumption.

Keywords: Mobile agent, Dynamic software composition, Intelligent robot control.

1   Introduction

In the last decade, robot systems have made rapid progress not only in their behaviors but also in the way they are controlled. In particular, control systems based on multiple software agents have been playing important roles. Multi-agent systems introduced modularity, reconfigurability and extensibility to control systems, which had been traditionally monolithic. It has made easier the development of control systems on distributed environments such as multi-robot systems. On the other hand, excessive interactions among agents in the multi-agent system may cause problems in the multiple robot environments.

In order to mitigate the problems of excessive communication, mobile agent methodologies have been developed for distributed environments. In a mobile agent system, each agent can actively migrate from one site to another site. Since a mobile agent can bring the necessary functionalities with it and perform its

N.T. Nguyen (Ed.): Transactions on CCI XI, LNCS 8065, pp. 119–136, 2013.
© Springer-Verlag Berlin Heidelberg 2013

tasks autonomously, it can reduce the necessity for interaction with other sites. In the minimal case, a mobile agent requires that the connection is established only when it performs migration [1].

The model of our system is a set of cooperative multiple mobile agents executing tasks for controlling a pool of multiple robots [2]. The property of inter-robot movement of the mobile agent contributes to the flexible and efficient use of the robot resources in addition to reducing the number of inter-communication as mentioned above. A mobile agent can migrate to the robot that is the most conveniently located to a given task, e.g. the closest robot to a physical object such as a soccer ball. Since the agent migration is much easier than the robot motion, the agent migration contributes to saving power consumption. Here, notice that any agents on a robot can be killed as soon as they finish their tasks. If the agent has a policy of choosing idle robots rather than busy ones in addition to the power-saving effect, it would result in more efficient use of robot resources.

In this paper, we show that the effectiveness of our control system in terms of resource usage and the efficiency of our system in terms of power consumption in two experimental setting. One setting consists of three physical robots cooperatively search targets [3] and the other setting consists of a larger number of virtual robots on a simulator also cooperatively search targets [4]. Since we implemented the simulator that reflects parameters given by the experiments with actual robots, we can expect the results of the simulation should be realistic enough. Although our example may look too simple, it could be extended to practical applications, or used as elements of them in a variety of applications because of its simplicity.

The structure of the balance of this paper is as follows. In the second section we describe the background. The third section describes the mobile agent class library that we have developed for controlling multiple robots. In our robot control system, the mobility that the software mobile agent system provides is the key feature that supports the ability of adding new functionalities to intelligent robots in action. The fourth section shows an example of intelligent robot systems in which robots search multiple target objects cooperatively. In the fifth section, we demonstrate how the properties of mobile agents can contribute to efficient use of robot resources through numerical experiments based on a simulator. Finally, we conclude in the sixth section.

2   Background

The traditional structure for the construction of intelligent robots is to make large, often monolithic, artificial intelligence software systems. The ALVINN autonomous driving system is one of the most successful such developments [5]. Putting intelligence into robots is, however, not an easy task. An intelligent robot that is able to work in the real world needs a large-scale knowledge base. The ALVINN system employs neural networks to acquire the knowledge semi-automatically [6]. One of the limitations of neural networks is that it assumes that the system is used in the same environment as that in which it was trained.

When the intelligent robot is expected to work in an unknown space or an extremely dynamic environment, it is not realistic to assume that the neural network is appropriately trained or can acquire additional knowledge with sufficient rapidity. Indeed, many intelligent robots lack a mechanism to adapt to a previously unknown environment.

On the other hand, multi-agent robotic systems are recently becoming popular in RoboCup or MIROSOT [7]. In traditional multi-agent systems, robots communicate with each other to achieve cooperative behaviors. The ALLIANCE architecture, developed in Oak Ridge National Laboratory, showed that cooperative intelligent systems could be achieved [8]. The architecture is, however, mainly designed to support self-adaptability. The robots in the system are expected to behave without external interference, and they show some intelligent behaviors. The observed intelligence, however, is limited due to the simple mechanism called *motivation*. Robots' behaviors are regulated by only two rules *robot impatience* and *robot acquiescence*. These rules are initially defined and do not evolve. In contrast, the goal of our system is to introduce intelligence and knowledge into the robots after they start to work [2]. Therefore, our system does not have any learning mechanism or knowledge acquiring mechanism. All the necessary knowledge is sent as mobile agents from other robots or the host computer.

An interesting research work of multi-agent robot control system was conducted at Tokyo University of Science [9]. Their work focused on the language aspect of robot control systems using multi-agents. They employed a hierarchical model of robot agents where the root agent indirectly manages all the agents. The lowest agents are physical devices and each has only one supervisor agent. Communications are performed through super-agent channels and sub-agent channels. Each robot has a hierarchically structured set of agents and the structure is rigidly constructed at the initial time. Therefore the structure of the control software is predetermined, and there is no concept of dynamic configuration of the structure of agents. The framework we present in this paper provides dynamic re-structuring of the set of agents and provides more flexibility in the real-world environments where any assumption cannot be expected.

For the communication aspect, they employ agent negotiation. In contrast, we employ agent migration so that our model can more suitably fit in a realistic multi-robot environment where the communication should be expected to be intermittent.

One notable feature of their system is the description language, called Multi-agent Robot Language (MRL). This language is based on the committed-choice concurrent logic programming language and compiled into the guarded Horn clauses [10,11]. This feature has advantages of transparency of the agent descriptions over our framework that is based on Java. The efficiency problem of logic programming is overcome by recompiling into C language. We also implement a descriptive language based on functional languages, Objective Caml and Scheme, in order to achieve the transparency of the agent descriptions [12,13].

The work most closely related to ours is the distributed Port-Based Adaptable Agent Architecture developed at Carnegie Mellon University [14]. The Port-Based

Agents (PBAs) are mobile software modules that have input ports and output ports. All PBAs have the map of the port addresses so that they can move other robots and combine themselves with other PBAs to compose larger modules. The usefulness of PBA architecture is demonstrated by the Millibot project also at Carnegie Mellon University [15]. In a robot mapping application, PBA is used to control the mapping robots, and when the working robot has a hardware failure, the PBA on the robot detects it and moves to an idle robot.

Software composition is clearly possible using port-based modules. The dynamic extension capability of our mobile agent control system, however, is another strategy for the composition of larger software.

The PBA is derived from the concept of port-based objects, designed for real-time control applications [16]. Therefore it may have advantages as a robot control mechanism. The framework we present in this paper is an exploration of the applications of mobile agents and software compositions through mobility and extensibility. Constructing robot control software by mobile agents and its dynamic extension is not only novel but also flexible due to the migration which agents perform autonomously. It may be superior for extensibility of working software.

3   Mobile Agent Controlling Robots

We assume that a mobile agent system consists of mobile agents and places. Places provide runtime environments, through which mobile agents achieve migration from one environment to other environments. When a mobile agent migrates to another place, not only the program code of the agent but also the state of the agent can be transferred to the destination. Once an agent arrives at another place through migration, it can communicate with other mobile agents on that place.

The mobile agent system we have used to control robots is based on an existing mobile agent system, called AgentSpace, developed by I. Satoh [17]. AgentSpace provides the basic framework for mobile agents. It is built on the Java virtual machine, and agents are supposed to be programmed in Java language. The behaviors of an agents on AgentSpace are determined by the following methods:

create: is called when initializing the agent,
destroy: is called when killing the agent,
leave: is called when migrating to another site, and
arrive: is called when arriving at the new site.

These methods are call-back methods which are invoked by a place i.e. a runtime environment of AgentSpace. The `create` or `destroy` method is called when the user requires to create or to kill mobile agents through GUI. The `arrive` and `leave` method are called in the process of migration. All the behaviors of an agent are determined by programs described in these methods. In the case where an agent needs to communicate with other agents, first, the agent requires the context of the current place that the agent resides. The context includes information of other mobile agents on the place, and therefore through that context, any agents can extract information such as the reference to a specific agent through its own name from the context. The reference can be used to invoke the methods of other agents on the same place. The context also provides the functionality of agent migration. The agent migration is achieved through method *move*. The method *move* receives an instance of URL class with IP and port number as an argument, and *move* the agent that is referred by the context.

We have extended AgentSpace and developed an agent library *Robot* that includes methods as shown by Fig. 1. In order to implement the methods, we have taken advantage of primitives of ERSP. ERSP is a software development kit with high-level interfaces tailored for controlling robots. These interfaces provide several high-level means for control such as driving wheels, detecting objects through a camera, checking obstacles through supersonic sensors, and intelligent navigations. They are written in C++, while mobile agents are described as Java classes that extend **Agent** class of AgentSpace. Therefore, we have designed *Robot* library that uses these interfaces through JNI (Java Native Interface). The library *Robot* has interfaces that are supposed to be implemented for the following methods:

initialize initializes flags for inputs from a camera and sensors,
walk makes a robot move straight within required distance,
turn makes a robot turn within required degree,
setObjectEvent resets the flag for object recognition with a camera,
setObstacleEvent resets the flag for supersonic sensors,
getObject checks the flag for object recognition,
getObstacle checks the flag for the sensors, and
terminate halts all the behaviors of a robot.

4   Robot Controller Agents for Target Searcher

In this section, we demonstrate that our model, which is a set of cooperative multiple mobile agents, is an efficient way to control multiple robots. In our robot control system, each mobile agent plays a role in the software that controls one robot, and is responsible to complete its own task. One agent with one specific task migrates to one specific robot to perform that task. In this manner, an agent can achieve several tasks one by one through the migration to robots one by one. This scheme provides more idle time for each robot, and allows other agents to use the idle robots for incomplete tasks. In that way, this scheme contributes in decreasing the total time of completing all the tasks. We will show these advantages in the numerical experiments.

```
package robot;

public class Robot {

    static {
        System.loadLibrary("RobotStatic");
    }

    static public native void initialize();
    static public native void walk(double distance, double speed,
        int wait, double timeOut);
    static public void walk(double distance, double speed, double timeOut) {
        walk(distance, speed, 500, timeOut);
    }
    static public native void turn(double angle, double speed, double timeOut);
    static public native void setObstacleEvent(int dir, double threshold);
    static public native void setObjectEvent(int objId, double threshold);
    static public native int getObstacle();
    static public native int getObject();
    static public native void terminate();
}
```

Fig. 1. Class library *Robot*

4.1   Controlling Robots

An intelligent multi-robot system is expected to work in a distributed environment where communication is relatively unstable and therefore where fully remote control is hard to achieve. Also we cannot expect that we know everything in the environment beforehand. Therefore intelligent robot control software needs to have the following features: 1) It should be autonomous to some extent. 2) It should be extensible to accommodate the working environment. 3) It should be replaceable while in action. Our mobile agents satisfy all these functional requirements.

Our control software consists of autonomous mobile agents. Once an agent migrates to a remote site, it requires minimal communication to the original site. Mobile agents can communicate with other agents on the same place so that the user can construct a larger system by migration to the place. The newly arrived agent can communicate with agents that reside in the system before its arrival, and achieve new functionality with them. If we find that the constructed software is not good enough to satisfy our requirements in a remote environment, we can replace the unsuitable component (an agent) with new component (an- other agent) by using agent migrations.

In the first experiment, we employed three wheeled mobile robots, which are called PIONEER 3-DX, as the platform for our prototype system. Each robot has two servo-motors with tires, one camera and sixteen sonic sensors. The power is supplied by rechargeable battery. Fig. 2 shows the team of robots in action for searching targets.

In the second experiment, we constructed a realistic simulator following information we have extracted from the observation of the behaviors of PIONEER 3-DX to show the scalability of our control system.

Fig. 2. A team of mobile robots are working under control of mobile agents

4.2   Searching a Target

Let us consider how to program a team of multiple robots to find a target. For such a task, the most straightforward solution would be to make all robots search for the target simultaneously. If the targets were comparatively fewer than the robots, however, most robots would move around in vain, consuming power without finding anything.

This problem can be more serious in our model where any robots can be shared by any agents, because the robots to which an agent with a new task is going to migrate may be already occupied by another agent with some different task. Especially, consider a case where the robots are working in an area where communications on wireless LAN are difficult. In such a case, even if one of the working robots finds the target, the other robot may not be able to know that fact. As a result, most robots continue to work to search that target in vain until time-out. Thus, this searching strategy could not only wastes the power but also increase the total costs of the multiple robots in aggregate. On the other hand, our strategy of using mobile agents achieves the suppression of the total cost due to the efficient use of idle resources as well as saving power consumption.

The core of our idea is finding the nearest robot to the target by using agent migration. Initially, an agent is dispatched from the host machine to a nearby

```
package agent.search;

import agentspace.*
import robot.Robot;

public class Search implements Agent, Mobile, Duplicatable, Preservable {

    public void arrive(AgentEvent evt, Context context) {
        //get ids of agents on the current place.
        AgentIdentifier[] aids = context.getAgents();

        if(aids.length == 1) {
            // If there is just itself on the current place,
            //execute  behavior() as the main behavior.
            behavior(context);
        }
        else {
            // If there are other agents on the current place,
            // it migrates to other robot.
            try {
                context.move(otherAddress());
            } catch(Exception e) { }
        }
    }


    public static void behavior(Context context) {
        while(true) {
            // It makes the robot rotate within 360 degrees.
            // If it finds a target, stop rotation and set the flag for detecting.
            Robot.turn(360.0, 5.0, 5.0);

            if (Robot.getObject().equals("TARGET1")) {
                // If detected thing is the target, it makes the robot go straight.
                Robot.walk(40.0, 3, 7.0);
            }
            else if (isExhausted()) {
                // If it has arrived at the last robot without finding anyting,
                // it makes the robot walk randomly.
                randomWalk();
            }
            else {
                // Otherwise, migrates to other robot.
                context.move (otherAddress());
            }
        }
    }
}
```

Fig. 3. *arrive()* method and *behavior()* method

robot in the multi-robots system. Then, the agent hops among the robots one by one and checks the robot's vision in order to locate the target until it reaches the robot that is the closest to the target. Until this stage, robots in the multi-robot system do not move; only the mobile agent migrates around so that robots can save power.

Once the agent finds the target, it migrates to the closest robot and makes the robot move toward the target. In our strategy, since only one robot dedicates to a particular task at a time, it is essentially similar to making each robot special for each task. Since the migration time is negligible compared to robot motion, our strategy is more efficient than such as we described before. If the agent visits all the robots without finding the target, the agent makes the last one move around randomly with wheels in order to find the target.

In our current multi-robot system, the robots' vision does not cover 360 degrees. Therefore a robot has to rotate to check its circumstance. Rotating a robot at its current position may capture the target and another robot closer to the target. Then the agent migrates to that more conveniently located robot. Notice that the rotation requires much less movement of the wheels than exploration, and it contributes to the power saving.

Details of our searching algorithm are the followings: 1) an agent chooses an arbitrary robot to which it migrates, and performs the migration, 2) as the agent arrives on the robot, it makes that robot rotate,where if the robot to which the agent migrates has been occupied by another agent, it migrates to another robot further,3) if the target is found, the agent makes the robot move to that direction; otherwise, goes back to step 1, and 4) at this stage, if all robots have been tried without finding the target, the agent makes the last robot do random-walk until it can find a target.

We have implemented this algorithm as an AgentSpace agent *search* as shown by Fig. 3. As soon as *search* agent has migrated to a robot, its *arrive()* method is invoked. *Arrive()* checks whether there are any other agents on the current robot or not. That can be achieved by checking agents' ids. This is achieved by calling *context.getAgents()*. If it finds only its own agent id, it means that there are no other agents. If it finds no other agents, it invokes *behavior()* as the main behavior of the robot. If it finds another agent id, it means the robot is occupied by the other agent, and the newly arrived agent immediately migrates to another robot in order to avoid interferences with the other agents.

The method *behavior()* first makes the robot rotate within 360 degrees to look around its circumstance. If it finds something, it stops the rotation of the robot, and sets the flag that indicates it detects an object. At this time, what is found can be checked through *Robot.getObject()*. As a result, if the return value corresponds to "TARGET1", it makes the robot go straight through *Robot.walk()*. Otherwise, it checks whether it has visited all the robots through *isExhausted()*. If there is no more robots to visit, it invokes *randomWalk()*, and makes the robot walk randomly in order to find the target at different angles. Otherwise, it migrates to one of the other robots.

5   Experimental Results

In order to demonstrate the effectiveness of our system, we have conducted numerical experiments on the example of target search that is mentioned in the previous section. In the experiments, we set a condition where robots search several targets, where searching each target corresponds to distinct task.

We have compared our approach based on mobile agents with other two strategies as follows:

EachForEach: allocates specific target to each robot. Each robot searches its own target, and does not search any other things as shown by Fig. 4(a), and

AllForEach: makes all robots move around for each target as shown by Fig. 4(b) which is the snapshot for searching a target.

The EachForEach approach is the simplest method. If there is no way to replace the program on a robot in action, this method would be used as the most realistic solution.

The AllForEach approach seems to consume the least time in order to search one target. In some case, it may also be the most efficient method for searching several targets. However, this approach may consume a lot of time in the area where the condition of connections among robots is changeable, because even if one robot finds a target, other robots may not be able to know the fact. In such a case, the task for searching one target might exhaust the time allowed for the entire team of robots. In our mobile agent based approach (Fig. 4(c)), a mobile agent is fixed to one robot at a time, and only one robot with the mobile agent searches one particular target until robots to which it migrates are exhausted. Eventually, the behavior becomes the same as the EachForEach approach.

5.1   Three Robots Experiment

First, in order to show that our approach is physically effective, we conducted experiments based on actual three robots in $3.5m \times 5.0m$ rectangle area which is the largest possible area where at least two of three robots have overlapping views.

We deal with the case where every target is the close to the different robot. In order to simulate realistic situations, we set several variations for some approach as follows:

For the EachForEach: a) robots for target A, B and C are respectively close to A, B and C , b) only robot for target A is close to A, and the other ones are close to different targets, and c) every robot is close to different target.

For the AllForEach: making all robots search one target at a time, and repeat for three targets. There is no other variation.

For mobile agent based approach: a) agents for target A, B and C initially migrate to robots close to A, B and C, b) some agents migrate one time after initial migration, and c) some agents migrate twice.

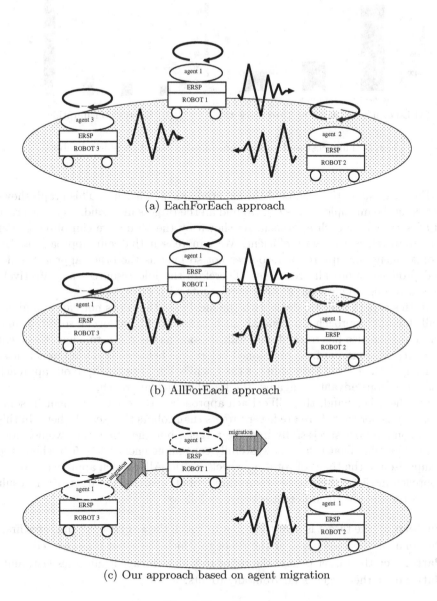

(a) EachForEach approach

(b) AllForEach approach

(c) Our approach based on agent migration

Fig. 4. Experiments

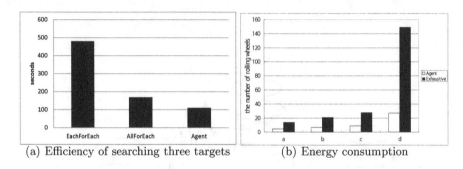

(a) Efficiency of searching three targets   (b) Energy consumption

Fig. 5. Time cost and energy consumption based on phisical robots

Fig. 5(a) shows the results of the experiments on efficiency. This graph shows how long the multiple robots take to find all the targets in seconds as the average of all settings for each approach. As shown by the figure, we can observe that our approach is the most efficient. We can reason that our approach makes mobile agents occupy robot resources not so long as the other approaches do, and produces more idle resources. Therefore the idle resources are effectively reused through migration of other new agents with other tasks.

In this experiment, we dealt with the ideal case that does not cause random-walk for our approach. In general, such cases would not occur so often, and our approach also has to randomly walk as soon as the robot to which the agent migrates are exhausted. As mentioned before, the behavior of our approach after visiting all the robots is the same as the EachForEach approach, so our approach has an obvious advantage against the EachForEach approach.

On the other hand, the AllForEach approach may be more efficient in some cases. Consider that there are fewer targets than robots that search them. In this case, some robots successfully find the targets, but the other ones would move around in vain. That results in wastefully consuming energy. Fig. 5(b) shows the comparison of the times of rotating wheels for the AllForEach approach and our approach in the extreme case where three robots search one target. Each result shows the time in the different settings as follows:

Pattern a: the target is near the robot to which the *search* agent migrates first,
Pattern b: the target is near the robot to which the agent migrates second,
Pattern c: the target is near the robot to which the agent migrates last, and
Pattern d: the target is far from any robots.

It is reasonable to assume that energy consumption of servo-motors is linear to the wheel rotation times. It is clearly observed that, in all the settings, our approach consumes less energy than the AllForEach approach. Furthermore, as mentioned before, the AllForEach approach can waste plenty of time in the case where the condition of connections among robots is changeable.

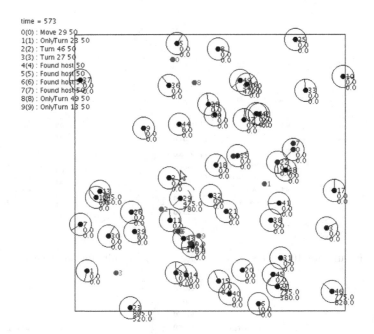

Fig. 6. A snapshot of a running simulator

5.2 Large Scale Simulation

Second, in order to demonstrate the effectiveness of our system in a large scale environment, we have implemented a simulator of the target search based on the real multi-robot system described in the previous section, and have conducted numerical experiments on it. On the simulator, moving and rotating speed of robots, and lags required in agent migration and object recognition are based on real values obtained in the previous experiments using PIONEER 3-DX with ERSP [3]. In the experiments, we set a condition where fifty robots are scattered in a 500×500 square field in the simulator, where searching each target corresponds to a distinct task. We have compared our approach based on mobile agents with other two strategies, AllForEach and EachForEach as well as the experiments with actual robots.

We have recorded the total moving distance and the total time of the robots that perform all the strategies. We have evaluated the results by changing the two parameters. They are the number of targets and the width of a view. The view means a circle with a robot as a center, and the robot can detect any objects in the circle as shown in Fig. 6. It is reasonable to assume that energy consumption of servomotors is linear to moving distance.

Bar charts of Fig. 7(a)–(d) show each of the total moving distances for 10, 30, 50, 70, and 90 targets respectively. The AllForEach strategy seems to achieve the least energy consumption over any number of targets. In some cases, it shows

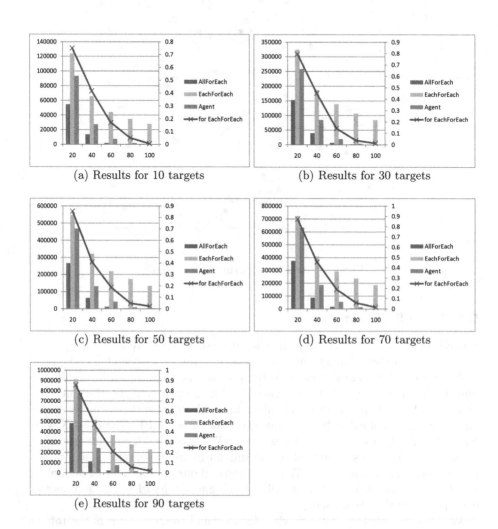

(a) Results for 10 targets

(b) Results for 30 targets

(c) Results for 50 targets

(d) Results for 70 targets

(e) Results for 90 targets

Fig. 7. The total moving distances for the view width: 20, 40, 60, 80, and 100

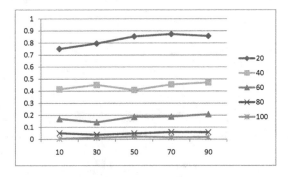

Fig. 8. The total moving distance over the numbers of targets

itself as the most efficient method for searching multiple targets. This approach, however, may consume a lot of energy when the condition of connections among robots is intermittent. Even though one robot finds a target, other robots may not be able to know the fact. In such a case, the task for searching one target might consume all the allowances for the entire team of robots. In our mobile agent based approach, on the other hand, a mobile agent is fixed on a certain robot after its migration, and as soon as the robot achieves the task, the entire multi-robot system turns to pursue the next object. As a result, the behavior of entire multi-robot system becomes effectively similar to the EachForEach strategy but it performs the same task just more efficiently.

The mobile agent system displays a remarkable saving of energy consumption compared to the EachForEach strategy. Furthermore, the more the width of a view increases, the more efficiency Agent gains than EachForEach, as shown by the line chart representing the ratio of the Agent to the EachForEach in Fig. 7(a)–(d).

Meanwhile, Fig. 8 shows the ratio of the total moving distance of the Agent strategy to the EachForEach strategy for each view width over the various numbers of targets. The flat lines illustrate the constant advantage of the Agent strategy over the EachForEach strategy regardless of the number of targets.

Fig. 9(a)–(d) shows the total time for searching out all the 10, 30, 50, 70, and 90 targets respectively. Since the total duration time is proportional to the total moving distance, the AllForEach strategy seems to be the most efficient among the three strategies. But it is not practical as mentioned above. On the other hand, the Agent strategy makes mobile agents occupy robot resources not so long as the other approaches do, and produces more idle resources. Then other new agents with other tasks can effectively use the idle resources through migration. We, however, observe that the Agent strategy shows less efficiency than that of the EachForEach where the width of a view is 20 shown in Fig. 9(b)-(d). In such cases, the Agent strategy often fails to find any target during the migration step due to the restricted view, and causes robots to randomly walk. We can conclude that the wider the view becomes, the more efficiently Agent works.

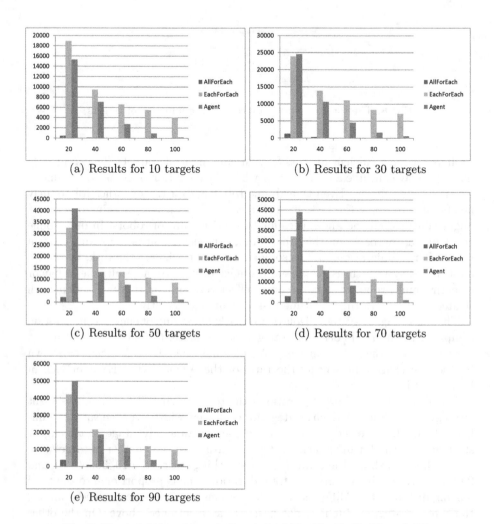

(a) Results for 10 targets

(b) Results for 30 targets

(c) Results for 50 targets

(d) Results for 70 targets

(e) Results for 90 targets

Fig. 9. The total time for each the view width: 20, 40, 60, 80, and 100

Thus, we believe that our approach is practical enough for controlling multiple robots in the real world in terms of the total cost and energy consumption.

6   Conclusions

We have presented a novel framework for controlling intelligent multiple robots. The framework helps users to construct intelligent robot control software by migration of mobile agents. Since the migrating agents can dynamically change the functionalities of the robot which they control, the control software can be flexibly assembled while they are running. Such a dynamically extending the control software by the migration of mobile agents enables us to make the base control software relatively simple, and to add functionalities one by one as we know the working environment. Thus we do not have to make the intelligent robot smart from the beginning or make the robot learn by itself. We can send intelligence later as new agents.

We have conducted experiments using three real robots. Through the experiments, we have successfully shown that our framework for controlling multiple robots can reduce energy consumption under the realistic circumstances. We also implemented a simulator that simulates the behaviors of a large scale team of cooperative search robots to show the effectiveness of our framework, and demonstrated that our framework contributes to suppressing the total cost of a multi-robot system in large scale cases. The numerical experiments show the volume of saved energy is significant. They demonstrate the superiority of our approach over more traditional non-agent based approaches.

Our future directions for research will include the addition of security features, refinement of the implementation of dynamic extension, additional proof of concept for dynamic addition of new functionality, and additional work on scheduling of conflicting tasks.

Acknowledgement. This work is supported in part by Japan Society for Promotion of Science (JSPS), with the basic research program (C) (No. 20510141), Grant-in-Aid for Scientific Research.

References

1. Binder, W.J., Hulaas, G., Villazon, A.: Portable resource control in the j-seal2 mobile agent system. In: Proceedings of International Conference on Autonomous Agents, pp. 222–223 (2001)
2. Kambayashi, Y., Takimoto, M.: Higher-order mobile agents for controlling intelligent robots. International Journal of Intelligent Information Technologies 1(2), 28–42 (2005)
3. Nagata, T., Takimoto, M., Kambayashi, Y.: Suppressing the total costs of executing tasks using mobile agents. In: Proceedings of the 42nd Hawaii International Conference on System Sciences. IEEE Computer Society CD-ROM (2009)

4. Abe, T., Takimoto, M., Kambayashi, Y.: Searching targets using mobile agents in a large scale multi-robot environment. In: O'Shea, J., Nguyen, N.T., Crockett, K., Howlett, R.J., Jain, L.C. (eds.) KES-AMSTA 2011. LNCS, vol. 6682, pp. 211–220. Springer, Heidelberg (2011)

5. Pomerleau, D.: Defense and civilian applications of the alvinn robot driving system. In: Proceedings of 1994 Government Microcircuit Applications Conference, pp. 358–362 (1994)

6. Pomerleau, D.: Alvinn: An autonomous land vehicle in a neural network. In: Advances in Neural Information Processing System 1, pp. 305–313. Morgan Kaufmann (1989)

7. Murphy, R.: Introduction to AI robotics. MIT Press, Cambridge (2000)

8. Parker, L.: Aliance: An architecture for fault tolerant multirobot cooperation. IEEE Transaction on Robotics and Automation 14(2), 220–240 (1998)

9. Nishiyama, H., Ohwada, H., Mizoguchi, F.: A multiagent robot language for communication and concurrency control. In: Proceedings of International Conference on Multi-Agent Systems, pp. 206–213 (1998)

10. Shapiro, E.: Concurrent Prolog: Collected Papers. MIT Press, Cambridge (1987)

11. Ueda, K.: Guarded Horn Clauses. PhD Thesis, University of Tokyo (1986)

12. Kambayashi, Y., Takimoto, M.: A functional language for mobile agents with dynamic extension. In: Negoita, M.G., Howlett, R.J., Jain, L.C. (eds.) KES 2004. LNCS (LNAI), vol. 3214, pp. 1010–1017. Springer, Heidelberg (2004)

13. Kambayashi, Y., Takimoto, M.: Scheme implementation of the functional language for mobile agents with dynamic extension. In: Proceedings of IEEE International Conference on Intelligent Engineering Systems, pp. 151–156 (2005)

14. Pham, T., Dixon, K.R., Jackson, J., Khosla, P.: Software systems facilitating self-adaptive control software. In: Proceedings of IEEE International Conference on Intelligent Robots and Systems, pp. 1094–1100 (2000)

15. Grabowski, R., Navarro-Serment, L., Paredis, C., Khosla, P.: Heterogeneous teams of modular robots for mapping and exploration. Autonomous Robots 8(3), 293–308 (2000)

16. Stewart, D., Khosla, P.: The chimera methodology: Designing dynamically reconfigurable and reusable real-time software using port-based objects. International Journal of Software Engineering and Knowledge Engineering 6(2), 249–277 (1996)

17. Satoh, I.: A mobile agent-based framework for active networks. In: Proceedings of IEEE System, Man and Cybernetics Conference, pp. 71–76 (1999)

Agent Based Optimisation of VoIP Communication

Drago Žagar[1] and Hrvoje Očevčić[2]

[1] University of Osijek, Faculty of Electrical Engineering
[2] Hypo Alpe-Adria-Bank d.d., Zagreb, Croatia
drago.zagar@etfos.hr, hrvoje.ocevcic@hypo-alpe-adria.hr

Abstract. An optimisation process implies procedures and actions for decreasing costs and making the system more efficient. This paper proposes a model for simple management and optimisation of VoIP quality of service. The model is proposed in framework of VoIP communication optimisation based on simple measurement information incorporated in agent architecture for VoIP QoS management. The model is based on adapted E-model for packet communication in which MOS values are assigned from objective measurements of QoS parameters. The large set of objective and subjective experimental QoS measurements is performed in order to asses the range of model applicability in operational network. The comparison of the experimental results of MOS values with calculated values of parameter R, as well as the mapping between them, gives a good base for QoS optimisation based on simple real-time measurements. According to agent based architecture the basic operating procedures for VoIP agent management system with assumed optimisation actions are proposed.

Keywords: VoIP, Quality of Service, delay, Mean Opinion Score (MOS), R factor, Agents.

1 Introduction

E-model is defined by ITU-T association in order to assess the quality of speech and connection of subjective and objective communication parameters. The output values of E-model are scalars, called R-values or R-factor. R factor can serve for calculation of subjective Mean Opinion Score (MOS) value [9]. The adjustment of E-model has been elaborated in many papers based on research of VoIP networks, stating a possibility of E-model adjustment, as well as with certain limitations appliance in VoIP systems [8][9][14][15][16].

This paper proposes the adapted E-model where R factor and the corresponding R values are described. According to ITU-T standard G.107 [8], the basic values of the majority of parameters are defined under the same network conditions. The use of predefined values of parameters in E-model gives R factor of 93.2.

The large set of objective and subjective experimental QoS measurements is performed in order to asses the range of model applicability in operational network.

N.T. Nguyen (Ed.): Transactions on CCI XI, LNCS 8065, pp. 137–154, 2013.

The experiment was carried out consisting of two parts. First part comprises experimental measurements in testing environment of objective quantitative parameters for VoIP QoS. This data was used for calculation of MOS values within the adjusted E-model. The second part comprises experimental measurement of subjective MOS values with listeners' participation, according to ITU-T P.800 [9]. It has been shown that regarding the constant terms in which the experiment was carried out, the proposed adjustment of E-model is applicable.

The practical result of the paper is proper correlation between the objective and subjective measurements as a base for construction of agent based proactive system for QoS optimisation in VoIP communication.

2   Theoretical Premises

The influence on quality of service in the wholesome network can be supported by two areas. The first area pertains the software support installed at the end computer (e.g. operational system). The second area pertains to the very network that transfers data from and to the end computers. This paper focuses mainly on the second area.

2.1   QoS Parameters in VoIP Networks

In order to make the transfer of voice data over IP network as a quality replacement for the standard phone service of voice transfer over PSTN network, the users should be provided at least by the same speech quality. Similarly to other real time services, IP telephony is very sensitive to delay and delay variations. The techniques for QoS support make it possible that voice packages have special treatment, necessary for achievement of the desired quality of service [3][4]. The ubiquitous QoS techniques imply following procedures:

- supporting the allocated bandwidth,
- reduction of package loss,
- avoiding and managing network congestion,
- shaping network traffic,
- strategies of traffic priorities in the network.

In Voice transmission over network infrastructure supporting transmission of packets, frames or cells, it is very important to supervise and control the components affecting the total delay. The received speech quality usually depends on many factors, including compression algorithms, transmission errors and packet losses, echo, delay and jitter. Also, along these QoS parameters, VoIP network designers should take into account how often the voice connections will be used, user QoS demands, as well as the basic user business activity [3][5].

2.2   Adapted E-Model for VoIP Communication

In many VoIP networks is very hard to find the proper connection between VoIP and classic telephony. The example could be dialling the number by the user in classic telephony redirected at the central office, coded and further processed as VoIP traffic. Therefore, the implementation of classic E-model could be questionable and of limited value. Many QoS parameters, required for calculation in E-model, are not measurable and therefore it is necessary to adapt the model by introducing some new parameters and assumptions. There are many papers considering adaptation of the current ITU-T E-model for application in characteristic environments [2][6].

The retention factor is a phenomenon of wrong subjective assessment caused by position of worsening quality in the sequence stream. If the streaming data are damaged in one of the initial packets, the influence of retention will not be the same as in case the damage is present at the end of the session. Therefore, the user would provide lower assessment because the lower level of the last received quality, which is psychological effect [10]. The retention effect is also considered in this paper as a effect of wrong users' assessments at the end of the session after adding the traffic worsening quality of service.

If the conditions for deployment of conversation between two speakers are constant, the ITU-T P.800 standard suggests the application of basic, implied values for the majority of parameters. If the basic values are listed for all other parameters, R factor becomes dependent on only one parameter - delay and therefore the delay is the parameter for comparison of measurement results. The adapted E-model is described by [6]:

R factor is presented as:

$$R = Ro - Is - Id - Ie + A \qquad (1)$$

Is - sum of all impairments occurring simultaneously with the voice signal.
Id - impairment factor representing all delays. It groups the talker echo, the listener echo and end to end delay.
Ie - represents impairments due to the nature of equipment
A - Advantage factor
Ro is the basic signal-noise ratio, which includes the noise sources e.g. the *circuit noise* and the *room noise*:

$$Ro = 15 - 1.5(SLR + No) \qquad (2)$$

The further elaboration of equations gives the form in which is possible to list the basic values [8]. By elaboration of E-model with use of values stated in Annex A6, the basic signal-noise ratio is:

$$Ro = 94.77.$$

The sum of all impairments Is can be shown as [8]:

$$Is = Iolr + Ist + Iq \qquad (3)$$

$Iolr$ - worsening factor caused by low OLR values
Ist - sidetone worsening factor
Iq - quantizing distortion

After including the basic values follows:

$$Is = 1.43.$$

Delay impairment, Id, is defined by following worsening factors:

$$Id = Idte + Idle + Idd \qquad (4)$$

$Idte$ - talker echo worsening factor
$Idle$ - listener echo worsening factor
Idd – absolute delay worsening factor

Using the basic values and assumptions based on ITU-T standards, we get:

$$Id = 0{,}15 + Idd \qquad (5)$$

And ultimately, R parameter is:

$$\begin{aligned}
R &= Ro - Is - Id - Ie + A = \\
&= 94{,}77 - 1{,}43 - (0{,}15 + Idd) - 0 + 0 = \\
&= 93{,}19 - Idd
\end{aligned} \qquad (6)$$

Where Idd depends on only one parameter of these measurement delay Ta,. For the delay less than 100ms, $Idd = 0$ [9] (i.e. basic value), and finally [1]:

$$R = 93{,}19.$$

2.3   Mapping of R Factor to MOS Value

The practical relevance of R-factor is in real time calculation, in line with measurement on specific communication line and equipment. By using correct modelling, the mapping between R factor and subjective MOS value can be defined with high accuracy.

Fig. 1. shows the functional relation between MOS value and R factor as user's satisfaction with the service. E-model defines the ideal case when R=100, and after defining the degrading values the R factor for specific communication is determinated.

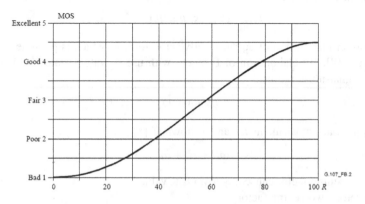

Fig. 1. The mapping between MOS value and R factor

R value can be mapped to MOS value with mathematical expression [8]:

$$MOS = \begin{cases} 1; & R < 0 \\ 1 + 0,035R + R(R-60)(100-R)*7*10^{-6}; & 0 < R < 100 \\ 4,5; & R \geq 100 \end{cases} \quad (7)$$

We can notice mathematical disproportion between R and MOS where MOS takes values less then 1 for 0<R<6.5. Therefore, the corrected relation is given by [8]:

$$MOS = \begin{cases} 1; & R \leq 6.5 \\ 1 + 0.035R + R(R-60)(100-R)*7*10^{-6}; & 6.5 < R < 100 \\ 4.5; & R \geq 100 \end{cases} \quad (8)$$

MOS value of 1 is defined for situations where R could be mathematically calculated to values less than zero. The standard also defines the formula for the calculation of R value from the known MOS value [8].

3   Experimental Model and Test Settings

The experimental test setting is shown in Fig. 2. The end system locations are connected by leased line with total bandwidth of 2Mbps. The VoIP access routers (Cisco 3640 and 2620) with voice modules are key traffic components of the settings, whose configuration is of great importance [18].

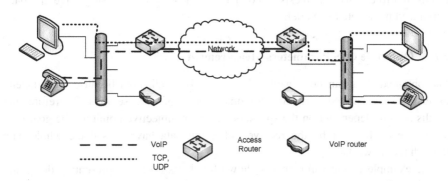

Fig. 2. Experimental test settings

For each of individual queuing, individual link bandwidth and individual codecs, the measurement process can be described in the following way:

The measurement is executed in a 100 seconds interval starting with VoIP traffic. After the first 30 seconds, TCP traffic is added, and after the following 30 seconds the additional UDP traffic is added. Fig. 3 shows the results of measurement scenario in which the time dependence of data amount and traffic type are presented.

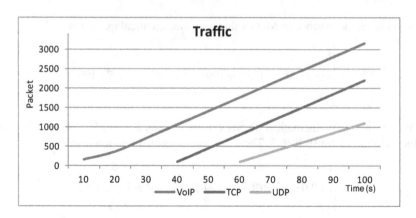

Fig. 3. VoIP, TCP and UDP traffic in measurement scenario

The basic approach of QoS assurance technique is to give priority to one type of traffic over another. Therefore, the performed techniques treated VoIP traffic as high priority traffic, while TCP and "normal" UDP traffic were in the same category – with no priority. The applied queuing techniques include FIFO, WRED, WFQ and LLQ.

TCP and UDP traffics are generated synthetically by the *hping* software [17]. It is important to emphasize that additional UDP traffic is not VoIP traffic type, and therefore it is not considered in the analysis. The settings of VoIP system must be identical in different measurement in order to assure consistency of measurement results and to confirm the thesis of comparison of objective and subjective approach introduced by adapted E-model.

3.1   Objective QoS Parameters Measurement

The obtained data can be considered as objective indicators because they depend exclusively on applied hardware equipment and software settings. Therefore, the results are not dependent on the speakers and their subjective opinion. The goal is to assess the quality felt by the receiver, after the data have passed the whole path through the network.

The example of measurements is shown by charts (Fig. 4.), presenting delay and jitter as well as the mean values for G.711 voice codec with total available bandwidth of 64 kbps. The results show that none of the tested queuing techniques provides satisfactory quality during the phases 2 and 3 where additional traffic was added. The delay as well as the jitter exceeds the maximal acceptable values resulting in low QoS. The only acceptable values are measured in phase 1 when only VoIP traffic is present. The best results were obtained using the codec G.711 in combination with WFQ queuing.

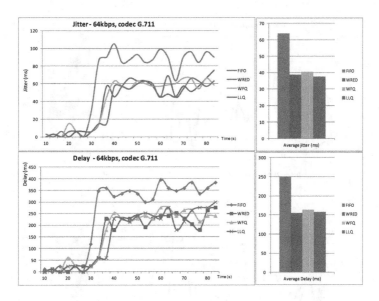

Fig. 4. Delay and jitter at 64kbps (G.711)

To compare the measurements with different parameters in VoIP communication Fig. 5. and Fig. 6. show the jitter and delay values for different total bandwidths (128 kbps and 2Mbps), using the same scenarios.

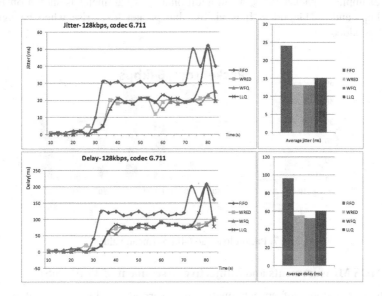

Fig. 5. Delay and jitter at 128kbps (G.711)

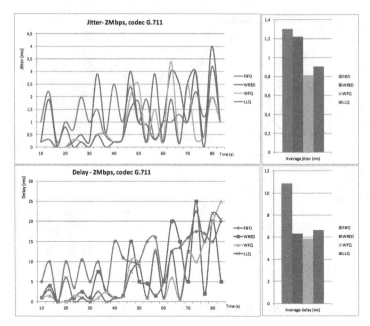

Fig. 6. Delay and jitter at 2Mbps (G.711)

Unlike, the results perceived in Fig. 6. (2 Mbps, G711) are acceptable in all measurements with WFQ as most appropriate queuing technique.

The example of packet loses, as an additional QoS parameter is shown on Fig. 7. for 64 kbps and G711 codec. It is visible that packet loses rise when background traffic was added.

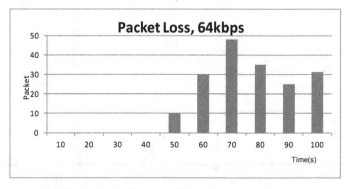

Fig. 7. Packet loses at 64kbps (mean values)

3.2   MOS Measurements and Subjective Assessment of VoIP QoS

Subjective assessment of speech quality was carried out by the model of Mean Opinion Score based on assessment given by 14 interlocutors. The test speech sequence was transferred through the network and "listened" at the other side, lasting for 100 seconds, similar as the previous measurement. The measurement parameters were set identically as in the previous measurement (bandwidth, codecs, traffic types and queuing). The text was reproduced from digital record, according to appendix A4.4, of the standard [9]. For the comparison to the first part of measurement providing objective assessments of the voice record, the additional TCP (the second record) and UDP traffic (the third record) were added.

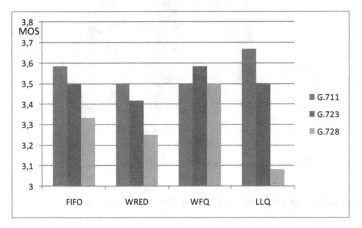

Fig. 8. The assessment of subjective quality at 64kbps

Fig. 8 shows the example of results of MOS subjective measurement for the same scenario and measurement presented by Fig. 4. Although, most of the obtained results doesn't meet required QoS, the best results were obtained by using G.711 codec in combination with LLQ queuing.

Fig. 9 shows the results of subjective VoIP quality assessment for scenario and measurement presented by Fig. 5. The obtained results are mostly satisfactory and the best results were obtained by using G.711 codec in combination with WRED and WFQ queuing.

The assessment of subjective quality at 2 Mbps is shown on Fig. 10. The perceived results fully meet the criteria of Quality of Service in VoIP communication and the best results were obtained for WFQ and WRED queuing technique.

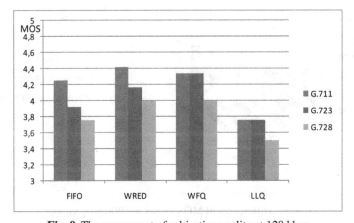

Fig. 9. The assessment of subjective quality at 128 kbps

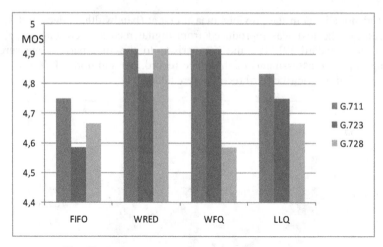

Fig. 10. The assessment of subjective quality at 2 Mbps

3.3   Subjective and Objective VoIP Quality Comparison

The objective measurements on the experimental model provided the input values for calculation of the corresponding MOS values. Under the same conditions the measurement on the network was carried out according to ITU-T P.800 [9]. The measured MOS values were provided by the listeners according to the standard terms. These values were compared in order to prove the regularity of the paper thesis.

The comparison indicates the largest difference after the first phase of VoIP communication, where the participants were listened the sound record transferred through the network with single VoIP traffic. After that, in the second part, the mixed traffic of VoIP and UDP was flowing through the network, and at the end also TCP

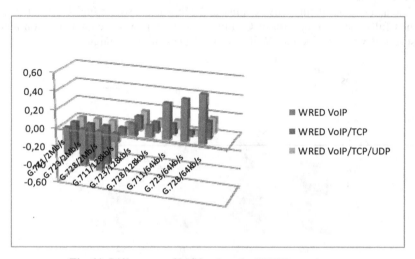

Fig. 11. Differences of MOS values for WRED queuing

traffic was added. The listeners assessed the quality with MOS values at the end of record when the influence of the congestion was the strongest. The example of results presenting the differences of the calculated and measured MOS values for WRED queuing mechanism are shown (Fig. 11).

The biggest difference appears at the bandwidth of 64kbps and by FIFO queuing. This results deviation is explained by the recency effect, which defines the psychological impact on the listener, where QoS worsening has a dominant effect by subjective measurement at the end of the session [14]. However, the proposed adaptive management system of VoIP communication quality is based on continuous measurement where the MOS values are calculated from the real time QoS parameters.

The calculation of subjective MOS values for VoIP communication could be the base for dynamic VoIP system QoS optimisation without expensive and complex equipment and measurement.

4   Agent Based VoIP Optimisation

Effective optimisation of VoIP quality of service could provide the users with optimal quality as well as the cost effective VoIP service. Since the subject of optimisation is the IP network as a transmission base of VoIP communication the optimised segments indirectly influence VoIP communication. The example of such optimisation is the incorporation of IDS/IPS (*Intrusion Detection System / Intrusion Protection System*) systems that eliminates or prevents additional types of traffic providing more resources allocated to VoIP communication. Such approach can generally be classified as the method of link optimisation where by managing available bandwidth we can assure the required service quality [12].

The way in which some service providers deal with VoIP optimisation is managing only one selected parameter that considerably affects the service quality. The examples include numerous codecs on VoIP market introducing solutions for bandwidth management and secondarily affecting the VoIP service quality [13].

Some commercial devices (e.g. SIP controllers) also provide service and network management as well as VoIP service optimisation [11].

This paper proposes the method for VoIP optimisation by management of service quality through adapted E-model. The reasons for selection of agent architecture for QoS management are the following:

- Simplicity of implementation: software agents upgrading without additional hardware implementation,
- Adjustability and flexibility: the agents activities are independent of environment, as well as of hardware and software implementations,
- Surveillance: it is possible to supervise and maintain the agent system easily and effectively,
- Price: the costs of implementation are reasonable; the total benefits of simple QoS management system are significant.

The use of subjective quality of service also has considerable advantages in comparison to objective approach. The following reasons can be stated as the major reasons for selection of subjective approach:

- Clarity and simplicity of service quality gradation at all communication sides, for service providers but also for service users,
- Compliance of results with objective measurement requiring more engagement, equipment and traffic,
- Optimized system for service quality support is a great target in communication service providing; it is also substantial to achieve cost benefit.

There are many commercial solutions, offering systems for continuous measurement of quality of service parameters of different Internet services [7][17]. Commercial solutions are usually expensive and often require significant system implementations. Monitoring itself is often the cause of delay as well as specific noise in network thus the goal is to come up with simple surveillance mode of network parameters and service quality management.

4.1   Agent Architecture for VoIP QoS Optimisation

The previous analysis represents the base for VoIP quality management system and therefore the optimisation model of VoIP communication is proposed as an agent based architecture. The proposal of agent based architecture for optimisation of VoIP communication is shown in Fig. 12. using the central agent management topology.

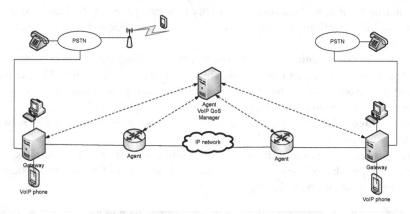

Fig. 12. Agent based architecture proposal

The distributed agents serve for communication and simple measurements, and continuously provide the information to VoIP QoS Manager Agent. The VoIP QoS Manager Agent manages session parameters according to the defined guidelines. Additionally, the information about the traffic type, bandwidth, codec, service type and clients in session is exchanged.

The basic guidelines can be initially defined and dynamically adjusted according to real time measurement results as well as good practices of VoIP communication. VoIP QoS Manager should also communicate with other system devices for sense of information exchange, as presented concisely in the Figure12.

The distributed agents can send real time information on interventions need while delivering the relevant measurement information. Many of this information already are generated by the components of the existing communication system. Fig. 13 shows the optimisation procedure of agent based VoIP communication.

According to Fig. 12 VoIP session can be established with different scenarios and between the clients of different types. Optimising and managing parameters of VoIP communication must also include the optimisation of these non VoIP sessions. Therefore, it is very important to define the guidelines for settings and behaviour of such system, while the delay monitoring is the easiest way to efficiently manage different communication circles.

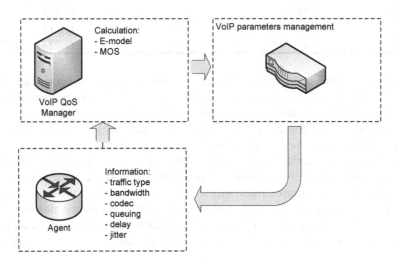

Fig. 13. VoIP Optimisation Procedure

The network quality of service changes dynamically because of variation of numerous parameters. To assure the user acceptable quality of service, the network critical parameters affecting the QoS should be carefully managed during the session. To achieve the desired optimised QoS in VoIP system the proposed agents should mutually exchange messages about actual network parameters. Some of them are already defined in VoIP signalisation while some must be obtained by measurement. According to above analysis delay is the most critical parameter which should be continuously measured and managed.

Table 1. Optimal codec combinations

Queuing	Bandwidth (kb/s)	VoIP/TCP/UDP	Optimal codec
FIFO	128	VT	G.722
FIFO	128	VTU	G.723
FIFO	64	VT	G.728
FIFO	64	VTU	G.728
WRED	128	VT	G.711
WRED	128	VTU	G.711
WRED	64	VT	G.711
WRED	64	VTU	G.711
WFQ	128	VT	G.711
WFQ	128	VTU	G.711
WFQ	64	VT	G.728
WFQ	64	VTU	G.711
LLQ	128	VT	G.711
LLQ	128	VTU	G.723
LLQ	64	VT	G.711
LLQ	64	VTU	G.723 / G.728

Additional techniques affecting QoS, such as optimization of packet size and delay variation compensation, can also be controlled by using the proposed architecture. The main advantage of this method is that a number of parameters and scenarios in which the communication circuit can be found are reduced to delay measurements followed by the basic defined parameters of the respective circuit.

The example of practical results of the experiments is shown in the Table 1. in form of optimal values of parameters for each tested scenario. The various scenarios are reduced to delay measurement with existent monitoring of the basic network parameters. By using the proposed architecture the access communication link is not relevant (Mobile, PSTN, WAN, LAN …) while at each session the same parameter is measured and the quality of service of VoIP communication is optimised in line with its amount.

The following example of pseudo-code sequence defines some agent actions based on experimental results.

```
Parameters:
PS=bandwidth              VP=traffic_type (V=VoIP,
RP=queue                  VTU=VoIP+TCP+UDP)
C=codec                   K=delay
'Before session:
If VP=VTU or PS<128 then
Set VP=V    'allow only VoIP traffic
Else
Set RP=LLQ   'best score is LLQ
Endif
```

```
'Agent loop:
If K>150ms then
  If PS=64 then
  Set VP=VoIP 'prioritization of traffic
EndIf
Else
      If PS<128 then
      Use cRTP    'use header compression
      Set C=G.723
      Optimize_packet_size 'packet size optimization
  Jitter_buffer ' jitter buffer optimization
  Else
Set VP=VoIP  'traffic prioritization
Endif
Endif
```

The implementation of software agents expands the range of possible locations where they are situated and if they are placed on the user side, they can monitor several parameters and optimise the system more precisely and efficiently.

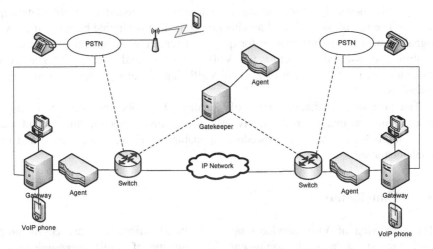

Fig. 14. The proposed agent architecture extended by VoIP Gatekeeper

The VoIP Gatekeeper is an optional component of the VoIP system which is used for switching and central managing of all endpoints of VoIP system. It manages the logical variables of proxy or gateway, and enables connections to PSTN networks with QoS mapping as well as the additional security settings. The Gatekeeper also provides address translation, bandwidth control and access to VoIP terminals and gateways.

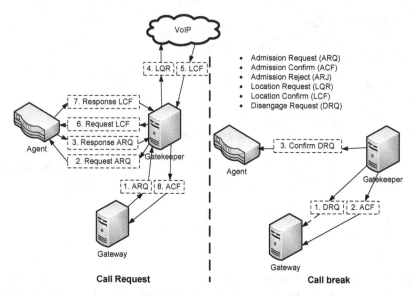

Fig. 15. Procedure of call signalization

Fig. 14. shows the proposed agent architecture extended by VoIP Gatekeeper device. The use of the extended architecture (Fig. 15.) is eligible because of network signaling with useful information for quality of service management (e.g. bandwidth control and admission control for VoIP terminals and gateways). The agents communicate over the IP network while VoIP Gatekeeper provides the management functions.

The proposed architecture for VoIP quality of service management is simple, flexible and adaptable, respectively. By using simple measurements and basic calculations for management procedures it enables efficient, cost effective and wide implementation.

5   Conclusion

The perception of VoIP service quality can be significantly improved by proper configuration of network equipment. The quality of VoIP communication is dependent on many parameters such as delay, delay variation, echo, as well as their interactions. By careful managing of parameters it is possible to significantly increase the user satisfaction and optimise VoIP communication even in extreme conditions.

In this paper we have proposed the adaptive agent based QoS management system by using subjective and objective parameters of VoIP communications. The proposed architecture enables dynamic optimization, based on simple and inexpensive mechanisms to ensure high performance VoIP communications. The most important measure of user satisfaction is a subjective quality rating. However, in dynamic real-time systems is nearly impossible, and certainly inefficient, provide continuous evaluation of subjective quality. Therefore, the linking objective and subjective QoS parameters is a crucial aspect of efficient quality management system. Consequently, the objective and subjective QoS parameters of VoIP communication in an experimental VoIP system were measured and analysed. The subjective quality evaluation was performed according to standard ITU-T P.800. The obtained results and performed analysis of subjective and objective assessments of service quality indicate that the mechanisms for ensuring quality services at the network level often can not provide a satisfactory quality of real-time services.

The adapted E-Model (ITU-T G.107) for linking objective and subjective assessments of service quality in voice communications was a base for estimation of the subjective QoS. The output value of the E-model is the R factor, which we can use to calculate the subjective MOS ratings. Based on the recommendations of the ITU-T G.107, when the VoIP users have the same conditions in the network we have used the basic values for most QoS parameters to calculate the R factor. One of the key points for successful optimization of VoIP communications, is to control the delay below some defined limit and therefore the proposed adapted E-model is dependent on only one variable parameter.

The main prerequisite for the cost effective and efficient QoS management is the implementation of minimal changes to the existing system. Therefore, this paper proposes an agent based architecture for VoIP QoS optimization. The proposed model uses the existing infrastructure upgraded by program agents for simple measurement of critical parameters as well as the effective QoS management. The experimental results can be a base for management guidance by using the adapted E-model.

.The model presented in this paper and the associated analyses can be applied as a general QoS management solution for real time applications, including different measurement scenarios, equipments, signalisation procedures, and service quality optimisation techniques, respectively.

Acknowledgments. This work is supported by the Croatian Ministry of Education, Science and Sports through the project 165-0362027-1479.

References

1. Black, U.: Internet Telephony: Call Processing Protocols. Prentice Hall (2001)
2. Walker, J.Q.: Assessing VoIP Call Quality using the E-model. netIQ Corporation (2001)
3. Understanding the Basic Networking Functions, Components, and Signaling Protocols in VoIP Networks, Part Number: 200087-002 (2007)
4. Mohd Nazri, I.: Analyzing of MOS and Codec Selection for Voice over IP Technology, Annals. Computer Science Series. 7th Tome 1st Fasc. (2009)
5. Advanced VoIP analysis and troubleshooting, Technical specification (2007), http://www.jdsu.com/test
6. Meddahi, A., Afifi, H.: Packet-E-Model: E-Model for VoIP quality evaluation. Computer Networks 50, 2659–2675 (2006)
7. Carvalho, L., Mota, E., Aguiar, R.: An E-Model Implementation for Speech Quality Evaluation in VoIP Systems. In: Proceedings of the 10th IEEE Symposium on Computers and Communications (2005)

8. The E-model, a computational model for use in transmission planning, ITU-T G.107, 1998, rev. (2000)
9. Methods for subjective determination of transmission quality, ITU-T P.800 (1996)
10. Huang, L.Y., Chen, Y.M., Chung, T.Y., Hsu, C.H.: Adaptive VoIP Service QoS Control based on Perceptual Speech Quality. In: The 9th International Conference on Advanced Communication Technology (February 2007)
11. The Need for SIP-Enabled Application Delivery Controllers, Radware, White Paper (June 2010)
12. Schluting, C.: WAN Optimization: Know Your Options (November 30, 2006)
13. Bradley, T.: Microsoft's Real-Time Codec (RTC) for VoIP optimization, SearchUnifiedCommunications.com
14. Clark, A.D.: Modeling the Effects of Burst Packet Loss and Recency on Subjective Voice Quality (2004)
15. Ding, L., Goubran, R.A.: Quality Prediction in VoIP Using the Extended E-Model. Carleton University 1125 Colonel By Drive, Ottawa, ON, K1S 5B6 (2003)
16. Matta, J., Pépin, C., Lashkari, K., Jain, R.: DoCoMo A Source and Channel Rate Adaptation Algorithm for AMR in VoIP Using the E model. Communications Laboratories, Inc., San Jose, CA 95110 (2003)
17. http://www.hping.org
18. Using Cisco Service Assurance Agent and Internetwork Performance Monitor to Manage Quality of Service in Voice over IP Networks, Cisco Systems, Document ID: 13938 (2009)

Towards Rule Interoperability:
Design of Drools Rule Bases Using the XTT2 Method*

Krzysztof Kaczor, Krzysztof Kluza, and Grzegorz J. Nalepa

AGH University of Science and Technology
al. A. Mickiewicza 30, 30-059 Krakow, Poland
{kk,kluza,gjn}@agh.edu.pl

Abstract. Despite the maturity of rule-based technologies and number of rule formalisms, the practical rule interoperability is still challenging. In a distributed environment where many knowledge engineers work in a collective way, this causes severe problems. This is a methodological paper, which introduces an approach that can be considered such an interoperability method. Its aim is to provide a unified and formalized method for knowledge interchange for the most common rule languages. Our approach involves three levels of interoperability abstraction: semantic, model and environment level. On each level different problems are addressed. In order to assess the appropriateness of such decomposition we provide a proof of concept solution concerning the interoperability between the Drools and XTT2 rule bases.[1]

1 Introduction

Rules are one of the most successful methods for declarative knowledge representation [16]. Due to their simplicity, they have been used for many years in different contexts and applications [13]. In the past, they were used mainly within RBS (*Rule-Based Systems*) [6]. In recent years, they have been applied in new fields like: BR (*Business Rules*) [48,15], Semantic Web [2,17] or CEP (*Complex Event Processing*) [14]. Currently, there are many languages that allow for encoding knowledge with rules e.g. Drools [5], CLIPS [13], Jess [12], OpenRules[2], ILog[3], etc.

Nowadays, the increasing significance of rule interchange (rule interoperability) can be observed. This is mainly due to the fact that distributed and collaborative environments for knowledge modelling and sharing became more common [35]. In such environments knowledge engineers work in a collective way [38], but often use different rule representation languages which can cause severe problems. The main purpose of an interchanging approach is to provide the means for reusing, publication and interchange of rules between different systems and tools. However, this is not a trivial task

---

* The paper is supported by the BIMLOQ Project funded from 2010–2012 resources for science as a research project.

[1] The paper extends concepts and preliminary results described in the paper [23] presented at ICCCI 2011 Conference in Gdynia, Poland.

[2] See: http://www.openrules.com

[3] See: http://www-01.ibm.com/software/websphere/ilog

N.T. Nguyen (Ed.): Transactions on CCI XI, LNCS 8065, pp. 155–175, 2013.

and requires taking several aspects into account. A complete interchange process should involve three levels of knowledge abstraction:

1. *Semantic level* – interchange process preserves the semantics of different rule bases.
2. *Model level* – takes the structure of the knowledge base into consideration.
3. *Environment level* – provides the support for design environments.

The rule interoperability method should assure support for all mentioned levels. These levels can be considered as the complete rule interoperability method, which provides a development pattern for proper interchange methods.

Nevertheless, the existing tools rarely provide support for knowledge interchange. Figure 1 shows the current support for rule interoperability in the most common rule-based systems. The meaning of the solid lines with arrows is straightforward. The solid lines indicate that the interchange methods are supported while the dashed lines mean they are neither supported nor exist.

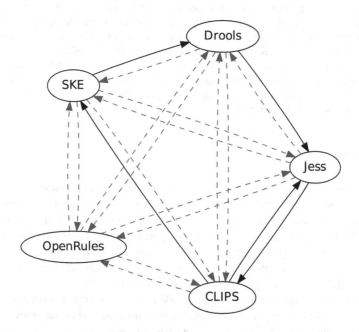

Fig. 1. Current rule interoperability methods

Our work focuses on a new approach to rule interoperability aiming at providing a unified and formalized methods for interchange. This methodological paper is an extension of the work described in [23] and presented at ICCCI 2011 Conference in Gdynia, Poland. The extension of this paper involves a broader look on the issues related to knowledge interchange. The concepts and preliminary results described in [23] constitute only a simple proof of concept for the problem discussed here. In comparison with the previous work, this paper provides new content of the state of the art and motivation sections. In addition, it discusses why the current rule interchange methods still involve many problems despite the maturity of rules. It gives a first glance on our approach to this matter and provides a prototype solution for rule interoperability method for SKE (*Semantic Knowledge Engineering*) [39] and Drools. The original contribution of this paper is the discussion and decomposition of the rule interoperability problem into three levels, where each of them involves different kinds of issues and problems concerning interchange process (see Section 2).

This paper is organized as follows: In Section 2, the motivation for our work is described. Section 3 discusses the existing approaches and technologies for rule interchange separately for each abstraction level. In Section 4, we describe focus of this paper, and then we present a proof of concept solution for Drools and XTT2 rule languages in Section 5. The paper is summarized and concluded with future work in Section 6.

2   Motivation

Use of rules involves important and still unsolved problems related to efficient representation of the rule-based knowledge (decision tables, trees) [49,31], agile development of the RBS, logical verification of the rule bases [29,8], etc. One of these problems is the lack of interoperability between different rule formats that causes difficulties in aligning rule bases to different existing systems and applications.

The lack of the efficient rule interoperability methods is the main motivation for our work. In particular, the following aspects confirm the need for solution of this problem:

– The rule-based technologies are an active field of research and development [50].
– The number of the rule-based applications is constantly increasing [14].
– Existing Rule-Based System shells [28] (e.g. CLIPS [13], Drools [5], OpenRules) provide different rule languages which are merely a programming solution [19].
– Lack of a commonly accepted formalized interoperability method for rules.
– Need for improvements of the practical knowledge maintenance with the common rule languages remains a challenge.

There are many existing approaches to these problems. Selected of them, suitable for comparison with our approach are discussed in Section 3. Our approach differs from the current methods which mainly constitute an intermediate interchange format. These methods are usually very abstract and complicated what makes the practical application very hard. Our approach does not aim at providing next intermediate format, but our goal is to create a unified rule representation formalism. Such formalism will provide logical representation for rule languages, even for those which are only programming solutions. Thanks to this, the language semantics can be unequivocally captured.

Nevertheless, the *semantic* level is only one of three levels that have to be taken into account in context of rule interoperability. In our work, we propose three levels of interchange method (see Figure 2). Only a method that takes all these levels into account can provide a proper rule interoperability. Omitting any of them can make practical application of the method impossible.

The interoperability on the *semantic* level has to take the semantics of the knowledge elements into account. This is why the rule language meaning must provide accurate definition. This can be assured by providing underlying logic which allows for

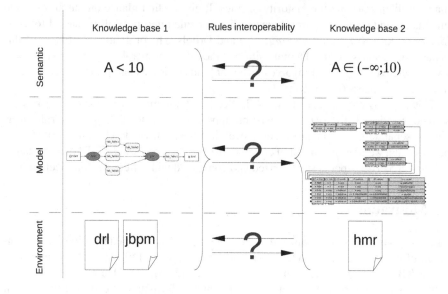

Fig. 2. Levels of complete rule interoperability method

unequivocal interpretation of the knowledge elements. Nevertheless, the existing rule languages are merely a programming solution which provide only well defined syntax. They rarely provide underlying logical interpretation with well defined semantics. Thus, such languages cannot be properly used for knowledge interchange because their semantics cannot be unequivocally translated.

As the semantics of the knowledge is the most important aspect in the interchange process, omitting this level may cause damage in the knowledge meaning [7]. For example: A rule language can restrict integer value of the attribute A to range $[-10, 10]$. In such case, semantics of the expression $A \leq 4$ can be interpreted as $A \in [-10, 4]$. The method, which do not take semantics level into account, may blindly translate this expression without changing its form. In some other language such expression can be interpreted as $A \in (-\infty, 4]$.

The *model* level involves issues related to knowledge base structure. Majority of the existing tools provide mechanisms for knowledge modularization. These mechanisms determine the knowledge base structure. The structured knowledge bases cannot be treated in the same way as unstructured (flat) ones. In a flat knowledge base, all rules are evaluated every time when the knowledge base is modified. Although the inference algorithms like Rete allow for avoiding iterating over the set of all rules, the pattern-matching network still involves all rules. This makes that each rule matched by Rete may be fired. In modularized rule bases, the knowledge structure may have impact on i.a. inference control. In Jess or CLIPS, the set of evaluated and fired rules is determined by both inference algorithm and modularization mechanism. Due to this fact, depending on the knowledge structure, one inference algorithm may produce different conclusions for knowledge bases containing semantically equivalent rules.

The *environment* interchange level is important in the context of practical method application. It concerns issues related to design environment such as support for rule language syntax, executing, tools integration and extension. Many of the existing solutions (see Section 3) provide very complex and general formalisms. Due to this fact, the practical application becomes hard. This level involves many technical issues related to importing, sharing and mapping knowledge into appropriate formats. The existing tools support different methods for knowledge storing e.g. databases, files (in different formats). A tool implementing interchange method has to provide support for retrieving and writing knowledge from/to these storage devices. The main problems that appear on this level and cannot be easily solved involve closed or undocumented file formats, lack of knowledge integration interfaces or closed tool architecture.

3   State of the Art in Rule Interchange

The previous section shows that rules interoperability methods have to involve many aspects. These aspects were grouped into three levels. This section provides the information concerning the existing technologies and solutions separately for each level.

Currently, there are several solutions that are being proposed for knowledge interchange. They provide formalism for rule representation, what allows for precise capturing of semantics. The most common are: RIF (*Rule Interchange Format*), RuleML (*Rule Markup Language*), R2ML (*REVERSE Rule Markup Language*), and KIF (*Knowledge Interchange Format*). Majority of them provide formalism and logical interpretation for rules. Nevertheless, they do not taking the other levels into account.

RIF [24] is the extensible rule interchange format for Semantic Web. It is indented to be an extensible interchange format for all rule languages. The architecture of RIF consists of several dialects, which are the XML-based rule languages with well-defined semantics. Some of these dialects, especially RIF Core, are at the early stage of development and can be superseded.

The practical application of RIF is limited because of its complexity and generic nature. Moreover, the RIF specification does not address several issues e.g. how to implement a transformation from a source rule language into RIF. Currently, the RIF support is developed in a number of tools e.g. RIFLE[4], SILK[5], fuxi[6], Eye[7], riftr[8]. Nevertheless, none of these tools provide full support for RIF.

RuleML [4] is an XML-based language for rule representation, which enables to express rules and modularize a knowledge base into stand-alone units. Moreover, it supports rule execution and exchange between different systems and tools. Each unit provides the support and semantics for a specific rule language and application. Thanks to the modular units, RuleML seems to be very flexible and extensible.

However, the RuleML representation does not provide any mechanisms for semantic evaluation of its elements. This is why the rule interchange can be inconsistent at

---

[4] See: http://sourceforge.net/apps/mediawiki/rifle

[5] See: http://silk.semwebcentral.org

[6] See: http://code.google.com/p/fuxi

[7] See: http://eulersharp.sourceforge.net/README#eye

[8] See: http://www.riftr.org

the semantic level. Furthermore, the provided units for specific rule languages lead to emergence of dialects. Thus, RuleML and RIF suffer from the same problem of complex dialects. The implementation of application that supports all the dialects becomes very hard. Hence, the tool support for this language is very weak. Moreover, it is not clearly specified how to interchange rule languages with complex data types and vocabularies, such as the Drools language, in RuleML.

A purpose of another rule markup, R2ML [51], is to capture rules formalized in different languages and interchange between rule formats and tools. R2ML allows for enriching ontologies by rules. It has an XML based concrete syntax validated by an XML Schema allowing for different semantics for rules. Rule concepts are defined with the help of MOF/UML, a subset of UML class modeling language. Later, the MOF/UML representation is mapped to the concrete markup syntax.

Because, R2ML does not provide any specific semantics, the accommodation of semantics of the target rule languages have to be defined every time. Furthermore, R2ML does not assure that defined interchange method for target rule languages is lossless.

KIF[9] is a computer-oriented language for the interchange of knowledge among different applications. It uses the first-order predicate logic to express arbitrary sentences. As KIF is logically comprehensive and provides declarative semantics, it is possible to understand the meaning of the expressions without usage of interpreter. Thus, KIF differs from other rule languages which are based on particular interpreters. Because the KIF is not as efficient as a specialized knowledge representation[10], it is not recommended to be used as an internal knowledge representation for applications.

KIF and their successors (SUO-KIF, LKIF) suffer from the same reasons as the SBVR (*Semantics of Business Vocabulary and Business Rules*) standard [45]. The specification of KIF is very complicated and vague. Similarly to SBVR, KIF provides very complex meta-model consisting of large number of classes and from the practical point of view, this causes that there is a lack of tools supporting KIF.

The mentioned technologies focus mainly on rules and their semantics. The current solutions do not provide support for issues related to *model* level. Nevertheless, Section 2 shows that *model* level is not meaningless and cannot be omitted.

The *model* level is mainly related to knowledge structure. So, this level provides a broader view on the interchange process because is not limited to rules as a basic knowledge unit but also takes relations between rules into account. The results of our research show that existing rule-based tools and technologies provide different structures of the knowledge bases [27].

In the simplest case, the knowledge base is flat i.e it is not structured. Such knowledge base can be easily interchanged on the *model* level because the structure does not have to be taken into account. This can be done even when another rule base provides a knowledge structure. However, the current tools like CLIPS, Jess, Drools, support mechanism for rules modularization.

CLIPS offers functionality for organizing rules into the so-called *modules*. They allow for limiting access to rules from other modules. They can be compared to global and local scopes in programming languages. The Rete algorithm implemented in CLIPS

---

[9] See: http://www.upv.es/sma/teoria/sma/kqml_kif/kif.pdf
[10] See: http://logic.stanford.edu/kif/introduction.html

builds separate pattern-matching network for each module. This affects inference control because only rules placed in the current module are evaluated and can be fired.

The Jess tool provides very similar modularization mechanism like CLIPS. This mechanism also affects inference control. The main difference between CLIPS and Jess modules is in the way how the Rete algorithm handles modules mechanism. In Jess, one pattern-matching network involving all modules is built. Due to this fact, all rules are evaluated independently from knowledge modularization. Nevertheless, only rules from the current module can be fired.

Drools, in turn, provides *ruleflow–groups* mechanism. Similarly to previous tools (especially CLIPS), this mechanism also affects inference control. However, the approach to knowledge modularization in Drools is slightly different. Ruleflow groups are formed by assigning an appropriate value to the rule `ruleflow-group` attribute. Moreover, Drools uses this mechanism for graphical representation and modeling of the knowledge structure and inference flow. Currently, this is done by jBPM workflow management system. In the past, this was supported by the Flow module.

The *environment* level of the interchange method has to identify the possibilities of method integration with tools. This integration can be done in several ways:

- by embeding tool API,
- by using tool communication protocol,
- by using plugin mechanism,
- by files supported by tools.

The integration based on files is the most common one. It requires to build a translator between file formats. Usually a translator is a separate application or script dedicated for this particular translation. Unfortunately, there are no technologies supporting translation between any formats. Only in some particular cases, implementation of translator can be supported by existing technologies e.g. XSLT – for translations from one XML-based format to another XML-based format [11,41]. However, many of the existing tools like CLIPS, Jess, Drools, OpenRules use other, hard to process, formats. Nevertheless, these tools also provide integration mechanisms: CLIPS API[11], Jess API[12], Drools API[13] and OpenRules API[14].

To summarize, the existing interchange methods involve mainly *semantic* level. Nevertheless, they are mostly too abstract and general and provide complex meta-models and vague formalisms, which cause the practical application hard. Furthermore, these methods are too narrow in the context of knowledge interoperability. They have to provide broader view on interchange process and take the issues related to *model* and *environment* level into account. This is why the current rule interchange methods suffer from the lack of the supportive tools and the formalism that can be used in practical applications. In fact, some rule interchange languages provide logical foundation, however in order to assure the semantically coherent rule interoperability, they can be reasonably

---

[11] See: http://clipsrules.sourceforge.net/documentation/v630/apg.

[12] See: http://www.jessrules.com/jess/docs/71/embedding.html.

[13] See: http://docs.jboss.org/drools/release/5.4.0.Beta2/droolsjbpm-integration-docs/html_single.

[14] See: http://openrules.com/docs/man_api.html.

applied only to rule languages that provide formalized semantics. This problem also limits the rule base maintenance methods because the different tools cannot share their features in rule base modeling.

4   Approach to Multi-level Rule Exchange

The omnipresent challenge in knowledge engineering is the lack of consistent rule interoperability methods. Nevertheless, the focus of this paper is on rule interoperability levels which are presented based on the example of interchange between XTT2 (*eXtended Tabular Trees version 2*) [44,33] and the Drools rule language [5]. This section provides a description of XTT2 and the Drools rule language on the three rule interoperability levels: semantic, model and environment levels.

4.1   XTT2 Rule Representation

Semantic level: Formalization of the knowledge base XTT2 is a hybrid rule representation in SKE (*Semantic Knowledge Engineering*) [39] methodology. It combines decision trees and decision tables forming a transparent and hierarchical visual representation of the decision tables linked into a decision network structure. It also provides a supportive conceptual method allowing for automated prototyping of rule base [42]. The name XTT2 was kept to provide compatibility with previous works.

XTT2 rule language is based on the ALSV(FD) (*Attributive Logic with Set of Values over Finite Domains*) logic [43,30]. This allows for precise and formal definition of all XTT2 knowledge elements as well as its semantics [44]. Contrary to the majority of other systems where a basic knowledge item is a single rule, in the XTT2 formalism a basic component displayed, edited and managed at a time is an *extended decision table*. Such a table is logically equivalent to a set of rules and can be considered as a modularization unit. Moreover, thanks to the ALSV(FD) logic, an expressiveness of XTT2 is greater than the classic (mostly propositional) rule languages, e.g. it allows for formal specification of non-atomic values in rule conditions, see [43] for more details. XTT2 rules can also be used combined with business processes [34].

Model Level: Inference Model. The XTT2-based models provide two levels of knowledge abstraction:

- the lower level – where a single knowledge component defined by a set of rules working in the same context is represented as a single XTT2 table, and
- the higher level – where the structure of the whole XTT2 knowledge base (consisting of XTT2 tables) is considered.

Such knowledge representation provides not only high density of knowledge visualization, but assures transparency and readability. The example of the visual representation of XTT2 table is depicted on Figure 3.

The main goals of the XTT2 representation are to provide an expressive formal logical calculus for rules as well as a compact, structural and visual knowledge representation and design. The representation allows for advanced inference control and formal

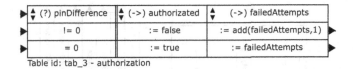

▸	♦ (?) pinDifference	♦ (->) authorizated	♦ (->) failedAttempts	
▸	!= 0	:= false	:= add(failedAttempts,1)	▸
▸	= 0	:= true	:= failedAttempts	▸

Table id: tab_3 - authorization

Fig. 3. An example of the XTT2 table

analysis of the production systems. The important advantage of the method is the internal structure of the XTT2 knowledge representation. As the rules that work together are grouped into contexts, the inference mechanism can work more efficient because only the rules from the focused context are evaluated. Moreover, the internal representation and structure makes the knowledge base suitable for visual editors.

An XTT2 knowledge base can be depicted as a graph. The tables constitute the graph nodes and correspond to the contexts in the knowledge base. The rules that belong to a specific context are placed in a single table in a network of tables. An exemplary XTT2 knowledge base graph for the cashpoint case study is presented in Fig. 4.

Environment Level: XTT2 Tool Support. Within the HeKatE project a number of tools supporting the visual design and implementation of the XTT2-based systems have been developed (see [22]). The two most important are HQEd and HeaRT.

HQEd (*HeKatE Qt Editor*) [21] is an editor for designing the XTT2-based expert systems. The editor supports visual modeling of rules and integration with the external systems. HQEd allows for designing a rule base in a visual way by using rules modularized in the network of decision tables. It also provides an interface for plugins which can extend its functionality and allow for integration with other systems. Moreover, HQEd provides two mechanisms that assure the high quality of the knowledge base:

– *Visual design* – visual representation makes the knowledge base more transparent what allows for more effective finding and fixing errors and anomalies.
– *Verification* – the design process is supported by two types of verification, which help discover the anomalies and errors that were omitted by designer.

HeaRT (*HeKatE Run Time*) [36] is a lightweight embeddable rule inference engine which uses XTT2 as rule specification. The XTT2 representation allows HeaRT to support two inference modes: DDI (*Data Driven inference*) and GDI (*Goal Driven inference*) which are forward and backward chaining algorithms. It is important to mention that HeaRT can be easily integrated with other solutions, e.g. Semantic Wiki [1], the Pellet reasoner [40].

4.2 Drools Project

Semantic Level: Drools Rule Language The semantics of the Drools rule language is not defined in a formal way and does not provide any logical interpretation. Drools rules

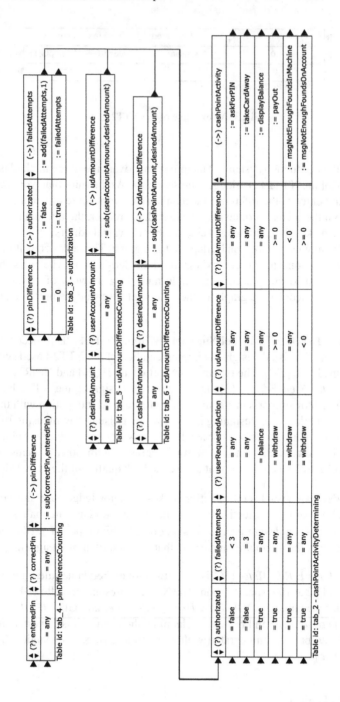

Fig. 4. An example of the XTT2 diagram for the cashpoint example

can be encoded in `drl`[15] format, which allows for declarative definition of the application logic. The knowledge base can be later easily imported into a JAVA application.

The Drools rule language is only a programming solution. It provides syntax for rules definition that consists of two parts: conditional (LHS) and decision (RHS). The general schema of rule is as follows:

```
rule "name"
     attributes
     when
         LHS
     then
         RHS
end
```

Rule attributes constitute simple (always optional) hints to how the rule should be treated. LHS is the conditional parts of the rule, which follows a certain syntax. RHS is basically a block that allows dialect specific semantic code to be executed.

Model Level: Drools Decision Tables and Workflow Model. In the case of Drools, the knowledge base is constructed as a workflow net. In the earlier versions, Drools Flow serves as a tool for designing the inference flow. In Flow the inference graph was simple and consists of special blocks and transitions that can be joined and split. Special blocks could be either single rules or ruleflow groups containing sets of rules.

Although ruleflow groups can be managed as decision tables, Drools does not provide any dedicated design tool for creating decision tables. Such decision tables for sets of rules having the same schema are simply created in MS Excel, OpenOffice or as CSV (*comma-separated values*) file (see Figure 5). Each rule, stored as a table row, consists of columns marked as "CONDITION" (LHS) and "ACTION" (RHS). Each attribute and logical operator takes one column header while table cells contain values corresponding to appropriate table headers. It is important to mention that during inference process, these decision tables are transformed into separate rules.

The workflow model in Drools used to be designed using the Drools Flow module, in which it was possible to design the inference flow in a visual way. There are also attempts to use some visual modeling languages for rules, such as URML or BPMN [32,46,10] and integrating them with Drools, however, these techniques allow for modeling only single rules and are not a part of the Drools project.

Environment Level: Drools and jBPM. With Drools 5.2, released in the middle of 2011, Drools Flow was replaced with jBPM5 module. JBoss jBPM [20] is a workflow management system and a platform for executable process languages ranging from business process management over workflow to service orchestration. The system architecture is based on the WfMC's reference model [18]. The jBPM project consists of several components:

---

[15] See: `http://docs.jboss.org/drools/release/5.4.0.Beta2/`
`drools-expert-docs/html_single/index.html#d0e2774`.

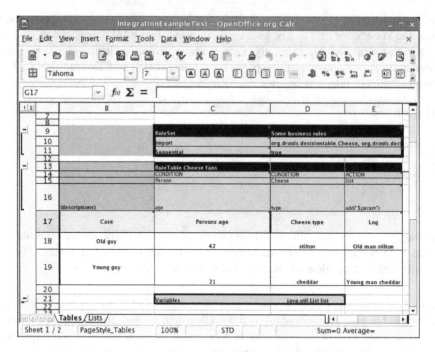

Fig. 5. Spreadsheet-based rules modeling

– the process engine for executing business processes and two optional core services (the history log for logging information about the current and previous state of all process instances and the human task service for the human task life cycle if human actors participate in the process),
– graphical editors for defining business processes (two types of editors are available: the Eclipse plugin and the web-based designer),
– the Guvnor repository for storing all the business processes,
– the jBPM console for managing business processes task list and reports.

jBPM has pluggable architecture and constitutes an extensible and customizable tool. This flexible and open-source BPMS (*Business Process Management Suite*) has been used for several workflow modeling purposes [3,47,52].

The tool allows for executing processes defined using the BPMN 2.0 XML format. However, the full BPMN 2.0 specification includes such details as choreographies and collaboration. The jBPM tools focus only on specifying executable processes [20]. Thus, in our research, only the subset of internal Business Process Model is considered.

5   Proposal of Rule Interchange Solution for XTT2 and Drools

In the previous section, we provided the description of XTT2 and Drools rule languages on the three rule interoperability levels: semantic, model and environment levels. In this section, we present a proof of concept solution concerning the cashpoint case study as a prototype method for rule interoperability between the two representations. This practical control system case study will be used to evaluate the discussed interoperability approach against the problems defined in Section 2.

The case study considered here is based on a cashpoint case described in [9]. The business logic of this case study designed in XTT2 was presented in Fig. 4. The detailed analysis of this simple, yet illustrative system, can be found at the webpage of the HeKatE project benchmark cases[16]: "A cashpoint is composed of a till which can access a central resource containing the detailed records of customers' bank accounts. A till is used by inserting a card and typing in a Personal Identification Number (PIN) which is encoded by the till and compared with a code stored on the card. After successfully identifying themselves to the system, customers may either make a cash withdrawal or ask for a balance of their account to be printed. Withdrawals are subject to a user resources, which means the total amount that user has on account. Another restriction is that a withdrawal amount may not be greater than the value of the till local stock. Tills may keep illegal cards, i.e. after three failed tests for the PIN."

Semantic Level Interoperability Solution. The XTT2 rule consist of condition and decision part. An example of the XTT2 rule from the CashPoint case is presented below:

```
xrule 'authorization/1':
     [pinDifference neq 0]
  ==>
     [authorizated set false,
      failedAttempts set (failedAttempts+1)]
  :'atm_action'.
```

This rule can be read as follows: *if the value of the attribute* pinDifference *is not equal to 0, then set value of attribute* authorizated *to* false *and increase value of attribute* failedAttempts *by 1.*

The first line defines rule (xrule) placed in the table authorization and labeled with 1. The third line contains a separator between conditional and decision parts of the rule. The last line provides the name of the next table that has to be evaluated. This line affects inference control and according to formalism is not a part of the rule.

The equivalent Drools rule can be written as follows:

```
rule "rule 01"
ruleflow-group "authorization"
lock-on-active true
    when
        $a : atm( getPinDifference() == 0 )
    then
        modify($a){
            setAuthorizated(false);
```

---

[16] http://ai.ia.agh.edu.pl/wiki/hekate:cases:
hekate_case_cashpoint:start

```
                    setFailedAttempts(getFailedAttempts()+1);
        }
end
```

The first line defines a rule label. In the second line, the assignment of the rule to `ruleflow-group` is defined. The `ruleflow-group` corresponds to the XTT2 table. In the XTT2 table, evaluation of each rule is done only one time. In order to assure the same behavior in Drools, a rule must define `lock-on-active` attribute. The meaning of the attribute is important whenever a `ruleflow-group` becomes active or receives focus. Any rule within that group has `lock-on-active` set to `true` and will not be activated any more; irrespective of the origin of the update, the activation of a matching rule is discarded. Only when the `ruleflow-group` is no longer active or loses the focus, those rules with `lock-on-active` set to `true` become eligible again for their activations to be placed onto the agenda[17]. The conditional part of the rule binds the object $a of `atm` type which fulfill the condition `getPinDifference() == 0`. In the decision part, this object is modified. This rule does not provide any information concerning inference control. The inference flow is defined in a visual way by a jBPM model.

Model Level Interoperability Solution. As a proof of concept for the interoperability on the model level, we prepared the translation rules between XTT2 and Drools format and developed DEPfH (*Drools Export Plugin for HQEd*) [23] which performs this nontrivial task. Although both Drools and XTT2 use rules as a knowledge representation, they store them in different formats. DEPfH by implementing translation between these two rule and knowledge base representations provides the rule interoperability on the model level [23].

During the DEPfH implementation, several differences between these representations had to be taken into consideration, such as the rule structure, decision tables structure, and links between the tables in the knowledge base.

In the case of decision tables, in Drools the logic operators are stored with attributes in table header and are common for all the column cells while in XTT2 a header of the table contains only attributes, common for all the rules in the table. When it comes to the decision table linking, in XTT2 each rule can have a link to another rule in the network of tables. These connections are used during the inference process and indicate a rule execution order. In Drools, rules in tables can not be connected directly. The connections are allowed only between ruleflow-groups, which can contain sets of rules or spreadsheet decision tables. It is worth mentioning that the Drools decision tables are used only during the design phase and they are not used during the inference process.

As Drools keeps the model elements in separate parts for the inference flow, decision tables and additional Java classes, the DEPfH plugin splits an XTT2 model into three files: one with the XTT2 model attributes, one with the decision tables, and additional file containing the inference flow. An example of the XTT2 model elements and their corresponding Drools elements are presented in Figure 6.

---

[17] See: `http://docs.jboss.org/drools/release/5.4.0.Beta2/drools-expert-docs/html_single/index.html#d0e2774`.

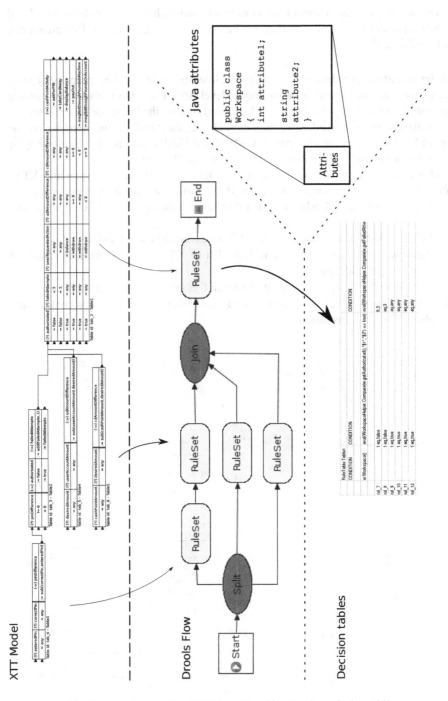

Fig. 6. An example of the XTT2 model and its Drools equivalent [23]

The deeper discussion about DEPfH implementation, limitations of both XTT2 and Drools representations, and how they can be overcome by using the DEPfH plugin can be found in [23].

Environment Level Solution. To achieve the interoperability on the environment level, rules should be interoperable on the model level first. Then, the tools should provide a possibility to exchange the rule knowledge and support the inference mechanism.

As jBPM provides the inference model for Drools, and HeaRT is the inference engine for XTT2, the goal of the integration of jBPM and HeaRT is to run the HeaRT engine in a server mode to handle communication from the jBPM client.

This is done by connecting the BPMN model tasks with the corresponding XTT2 tables. Communication is initialized from jBPM during the execution of the process by sending the appropriate IP address and port of submission in the HeaRT defined format:

1. *Rule task* with names starting from "H" are executing using HeaRT.
2. *Ruleflow groups* are associated with the list of XTT2 tables (optionally also with states) in the rule base.
3. If the model state is not specified, the HeaRT engine starts from its current state.
4. State attributes which are results of the HeaRT inference are stored as context variables of the process instances.

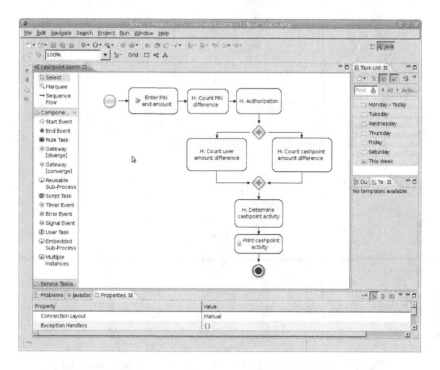

Fig. 7. An example of BPMN model for the CashPoint case study in jBPM

As a proof of concept for the translation and integration of jBPM and XTT2, we modeled the CashPoint example within the jBPM framework. For communication purposes the HConnect class was implemented. It uses the JHeroic library and allows for communication with the HeaRT rule engine, inference and processing the results along with the modeled workflow.

Figure 7 shows an example of the process model, which demonstrates the dependencies between rule tasks and XTT2 tables and the script task and the HeaRT output. Thanks to this, the model constitutes a visual inference specification for the rule base [27]. This is a part of our research concerning Business Processes and Business Rules integration [26,25]

6 Summary

The original contribution of this paper is the discussion and decomposition of the rule interoperability problem into three levels: *semantic*, *model* and *environment*. Each of these levels involves different kinds of issues and problems concerning interchange process (see Section 2):

1. *Semantic* level involves problems related to preserving knowledge semantics during interchange process.
2. *Model* level forces a broader view on interchange process because is not limited to rules as a basic knowledge unit but also takes relations between rules into account. A knowledge structure may impact on knowledge processing, what should not be omitted in an interchange method.
3. *Environment* level is related to interchange technical aspects. This involves issues and problems of integration.

Despite the maturity of Rule-Based Systems, the rule interoperability problem still exists. In general, the problem involves the following specific aspects:

- Lack of the logical interpretation of rule languages – this problem is related to the *semantic* level.
- Lack of formalized way for knowledge exchanging between rule representations – this issue can be observed on the *semantic* and *model* levels.
- Limited rule base maintenance methods – this problem appears on the *model* and *environment* levels.

Lack of well-defined rule interoperability solutions is a major challenge in a distributed environment where many knowledge engineers work in a collective way [37].

The focus of this paper is on the rule interoperability levels presented based on the example of interchange between XTT2 and Drools rule language. Evaluation of the provided proof of concept can be summarized as follows. On the *semantic* level the knowledge semantics is preserved during translation. The semantically equivalent knowledge elements were identified in informal way. On this stage of research any rule interchange formalisms cannot be provided. Then, the DEPfH translator takes the *model* level into account and maps an XTT2 model structure to the corresponding structure in Drools.

The rules from each XTT2 table are placed into corresponding ruleflow-groups. Moreover, The DEPfH integrates the interchange method with XTT2 and Drools on the *environment* level as well. The input knowledge for translator is retrieved from the HQEd plugin mechanism. On the output, this translator generates appropriate Drools knowledge, jBPM and HeaRT files.

The work presented in this paper is a methodological approach to knowledge interoperability. The main goal of our future work is to provide a formalized method for rule interoperability for the most common rule languages: CLIPS, Jess, Drools, OpenRules and SKE. The proposed method will be based on the ALSV(FD) logic and will allow for unified knowledge interoperability in the context of the mentioned rule languages. The formalized rule representation will be adapted to rule languages and allow for having a unified logical interpretation of the rule languages. The formulation of the unified rule interoperability format will facilitate capturing of the rule languages semantics in an unequivocal way. This allows for semantically coherent knowledge interoperability for considered rule languages. Next, the developed method will be analyzed in the context of its application or extension towards another rule languages.

References

1. Adrian, W.T., Bobek, S., Nalepa, G.J., Kaczor, K., Kluza, K.: How to reason by HeaRT in a semantic knowledge-based wiki. In: Proceedings of the 23rd IEEE International Conference on Tools with Artificial Intelligence, ICTAI 2011, Boca Raton, Florida, USA, pp. 438–441 (November 2011)
2. Antoniou, G., van Harmelen, F.: A Semantic Web Primer. The MIT Press (2008)
3. Bing, H., Dan-Mei, X.: Research and design of document flow model based on JBPM workflow engine. In: Proceedings from International Forum on Computer Science-Technology and Applications, IFCSTA 2009, vol. 1, pp. 336–339 (December 2009)
4. Boley, H., Tabet, S., Wagner, G.: Design rationale for ruleml: A markup language for semantic web rules. In: Cruz, I.F., Decker, S., Euzenat, J., McGuinness, D.L. (eds.) SWWS, pp. 381–401 (2001)
5. Browne, P.: JBoss Drools Business Rules. Packt Publishing (2009)
6. Buchanan, B.G., Shortliffe, E.H. (eds.): Rule-Based Expert Systems. Addison-Wesley Publishing Company, Reading (1985)
7. Cañadas, J., Palma, J., Túnez, S.: Defining the semantics of rule-based web applications through model-driven development. International Journal of Applied Mathematics and Computer Science 21(1), 41–55 (2011)
8. Coenen, F., et al.: Validation and verification of knowledge-based systems: report on eurovav99. The Knowledge Engineering Review 15(2), 187–196 (2000)
9. Denvir, T., Oliveira, J., Plat, N.: The Cash-Point (ATM) 'Problem'. Formal Aspects of Computing 12(4), 211–215 (2000)
10. Di Bona, D., Lo Re, G., Aiello, G., Tamburo, A., Alessi, M.: A methodology for graphical modeling of business rules. In: 5th UKSim European Symposium on Computer Modeling and Simulation (EMS), pp. 102–106 (November 2011)
11. Fong, J., Shiu, H., Wong, J.: Methodology for data conversion from XML documents to relations using Extensible Stylesheet Language Transformation. International Journal of Software Engineering and Knowledge Engineering 19(2), 249–281 (2009)
12. Friedman-Hill, E.: Jess in Action, Rule Based Systems in Java. Manning (2003)

13. Giarratano, J., Riley, G.: Expert Systems. Principles and Programming. Thomson Course Technology, 4th edn., Boston, MA, United States (2005) ISBN 0-534-38447-1

14. Giurca, A., Gašević, D., Taveter, K. (eds.): Handbook of Research on Emerging Rule-Based Languages and Technologies: Open Solutions and Approaches. Information Science Reference, Hershey (2009)

15. von Halle, B.: Business Rules Applied: Building Better Systems Using the Business Rules Approach. Wiley (2001)

16. Hendler, J., van Harmelen, F.: The Semantic Web: Webizing Knowledge Representation. In: Handbook of Knowledge Representation. Elsevier, New York (2008)

17. Hitzler, P., Krötzsch, M., Rudolph, S.: Foundations of Semantic Web Technologies. Chapman & Hall/CRC (2009)

18. Hollingsworth, D.: The workflow reference model. Issue 1.1 TC00-1003, Workflow Management Coalition (January 1995)

19. Jackson, P.: Introduction to Expert Systems. Addison–Wesley, 3rd edn. (1999) ISBN 0-201-87686-8

20. The jBPM team of JBoss Community: jBPM User Guide, 5.2.0.final edn. (December 2011), http://docs.jboss.org/jbpm/v5.2/userguide/

21. Kaczor, K., Nalepa, G.J.: Extensible design and verification enviroment for XTT rule bases. In: Tadeusiewicz, R., Ligęza, A., Mitkowski, W., Szymkat, M. (eds.) CMS 2009: Computer Methods and Systems: 7th Conference, Kraków, Poland, November 26-27, pp. 99–104. AGH University of Science and Technology, Oprogramowanie Naukowo-Techniczne, Cracow (2009)

22. Kaczor, K., Nalepa, G.J.: HaDEs – presentation of the HeKatE design environment. In: Baumeister, J., Nalepa, G.J. (eds.) 5th Workshop on Knowledge Engineering and Software Engineering (KESE 2009) at the 32nd German conference on Artificial Intelligence, Paderborn, Germany, pp. 57–62 (September 15, 2009)

23. Kaczor, K., Nalepa, G.J., Łysik, Ł., Kluza, K.: Visual design of Drools rule bases using the XTT2 method. In: Katarzyniak, R., Chiu, T.-F., Hong, C.-F., Nguyen, N.T. (eds.) Semantic Methods for Knowledge Management and Communication. SCI, vol. 381, pp. 57–66. Springer, Heidelberg (2011), http://www.springerlink.com/content/h544g4238716m320/

24. Kifer, M., Boley, H.: RIF overview. W3C working draft, W3C (October 2009), http://www.w3.org/TR/rif-overview

25. Kluza, K., Kaczor, K., Nalepa, G.J.: Enriching business processes with rules using the Oryx BPMN editor. In: Rutkowski, L., Korytkowski, M., Scherer, R., Tadeusiewicz, R., Zadeh, L.A., Zurada, J.M. (eds.) ICAISC 2012, Part II. LNCS, vol. 7268, pp. 573–581. Springer, Heidelberg (2012), http://www.springerlink.com/content/u654r0m56882np77/

26. Kluza, K., Maślanka, T., Nalepa, G.J., Ligęza, A.: Proposal of representing BPMN diagrams with XTT2-based business rules. In: Brazier, F.M., Nieuwenhuis, K., Pavlin, G., Warnier, M., Badica, C. (eds.) Intelligent Distributed Computing V. SCI, vol. 382, pp. 243–248. Springer, Heidelberg (2011)

27. Kluza, K., Nalepa, G.J., Łysik, Ł.: Visual inference specification methods for modularized rulebases. Overview and integration proposal. In: Nalepa, G.J., Baumeister, J. (eds.) Proceedings of the 6th Workshop on Knowledge Engineering and Software Engineering (KESE6) at the 33rd German Conference on Artificial Intelligence, Karlsruhe, Germany, pp. 6–17 (September 21, 2010), http://ceur-ws.org/Vol-636/

28. Liebowitz, J. (ed.): The Handbook of Applied Expert Systems. CRC Press, Boca Raton (1998)

29. Ligęza, A.: Intelligent data and knowledge analysis and verification; towards a taxonomy of specific problems. In: Ligęza, A. (ed.) Validation and Verification of Knowledge Based Systems: Theory, Tools and Practice, pp. 313–325. Kluwer Academic Publishers, Boston (1999)

30. Ligęza, A., Nalepa, G.J.: A study of methodological issues in design and development of rule-based systems: proposal of a new approach. Wiley Interdisciplinary Reviews: Data Mining and Knowledge Discovery 1(2), 117–137 (2011),
http://onlinelibrary.wiley.com/doi/10.1002/widm.11/pdf

31. Ligęza, A., Szpyrka, M.: Reduction of tabular systems. In: Rutkowski, L., Siekmann, J.H., Tadeusiewicz, R., Zadeh, L.A. (eds.) ICAISC 2004. LNCS (LNAI), vol. 3070, pp. 903–908. Springer, Heidelberg (2004)

32. Lukichev, S., Wagner, G.: Visual rules modeling. In: Virbitskaite, I., Voronkov, A. (eds.) PSI 2006. LNCS, vol. 4378, pp. 467–473. Springer, Heidelberg (2007)

33. Nalepa, G., Ligęza, A., Kaczor, K.: Overview of knowledge formalization with XTT2 rules. In: Bassiliades, N., Governatori, G., Paschke, A. (eds.) RuleML 2011 - Europe. LNCS, vol. 6826, pp. 329–336. Springer, Heidelberg (2011)

34. Nalepa, G.J.: Proposal of business process and rules modeling with the XTT method. In: Negru, V., et al. (eds.) Symbolic and Numeric Algorithms for Scientific Computing, SYNASC Ninth International Symposium, September 26-29, pp. 500–506. IEEE Computer Society, IEEE, CPS Conference Publishing Service, Los Alamitos, California, Washington, Tokyo (2007)

35. Nalepa, G.J.: PlWiki – a generic semantic wiki architecture. In: Nguyen, N.T., Kowalczyk, R., Chen, S.-M. (eds.) ICCCI 2009. LNCS, vol. 5796, pp. 345–356. Springer, Heidelberg (2009)

36. Nalepa, G.J.: Architecture of the HeaRT hybrid rule engine. In: Rutkowski, L., Scherer, R., Tadeusiewicz, R., Zadeh, L.A., Zurada, J.M. (eds.) ICAISC 2010, Part II. LNCS, vol. 6114, pp. 598–605. Springer, Heidelberg (2010)

37. Nalepa, G.J.: Collective knowledge engineering with semantic wikis. Journal of Universal Computer Science 16(7), 1006–1023 (2010),
http://www.jucs.org/jucs_16_7/
collective_knowledge_engineering_with

38. Nalepa, G.J.: Loki – semantic wiki with logical knowledge representation. In: Nguyen, N.T. (ed.) TCCI III 2011. LNCS, vol. 6560, pp. 96–114. Springer, Heidelberg (2011),
http://www.springerlink.com/content/y91w134g03344376/

39. Nalepa, G.J.: Semantic Knowledge Engineering. A Rule-Based Approach. Wydawnictwa AGH, Kraków (2011)

40. Nalepa, G.J., Furmańska, W.T.: Pellet-HeaRT – proposal of an architecture for ontology systems with rules. In: Dillmann, R., Beyerer, J., Hanebeck, U.D., Schultz, T. (eds.) KI 2010. LNCS, vol. 6359, pp. 143–150. Springer, Heidelberg (2010),
http://www.springerlink.com/content/r46p8m40432n7342/

41. Nalepa, G.J., Kluza, K.: UML representation for rule-based application models with XTT2-based business rules. International Journal of Software Engineering and Knowledge Engineering (IJSEKE) 22(4), 485–524 (2012)

42. Nalepa, G.J., Ligęza, A.: Conceptual Modelling and Automated Implementation of Rule-Based Systems. In: Software engineering: evolution and emerging technologies, Frontiers in Artificial Intelligence and Applications, vol. 130, pp. 330–340. IOS Press, Amsterdam (2005)

43. Nalepa, G.J., Ligęza, A.: HeKatE methodology, hybrid engineering of intelligent systems. International Journal of Applied Mathematics and Computer Science 20(1), 35–53 (2010)

44. Nalepa, G.J., Ligęza, A., Kaczor, K.: Formalization and modeling of rules using the XTT2 method. International Journal on Artificial Intelligence Tools 20(6), 1107–1125 (2011)

45. Semantics, O.M.G.: of Business Vocabulary and Business Rules (SBVR). Tech. Rep. dtc/06-03-02, Object Management Group (2006)
46. Pascalau, E., Giurca, A.: Can URML model successfully drools rules? In: Giurca, A., Analyti, A., Wagner, G. (eds.) ECAI 2008: 18th European Conference on Artificial Intelligence: 2nd East European Workshop on Rule-Based Applications, RuleApps 2008, July 22, pp. 19–23. University of Patras, Patras (2008)
47. Peng, L., Zhou, B.: Research on workflow patterns based on jBPM and jPDL. In: Proceedings from IEEE Pacific-Asia Workshop on Computational Intelligence and Industrial Application, PACIIA 2008, vol. 2, pp. 838–843. IEEE (December 2008)
48. Ross, R.G.: Principles of the Business Rule Approach, 1st edn. Addison-Wesley Professional (2003)
49. Szpyrka, M., Szmuc, T.: Decision tables in petri net models. In: Kryszkiewicz, M., Peters, J.F., Rybiński, H., Skowron, A. (eds.) RSEISP 2007. LNCS (LNAI), vol. 4585, pp. 648–657. Springer, Heidelberg (2007)
50. Tadeusiewicz, R.: Introduction to intelligent systems. In: Wilamowski, B.M., Irwin, J.D. (eds.) Intelligent Systems, 2nd edn. The Electrical Engineering Handbook Series. The Industrial Electronics Handbook, pp. 1-1–1-12. CRC Press Taylor & Francis Group, Boca Raton (2011)
51. Wagner, G., Giurca, A.: R2ml: A general approach for marking up rules. In: Bry, F., Fages, F., Marchiori, M., Ohlbach, H. (eds.) Principles and Practices of Semantic Web Reasoning, Dagstuhl Seminar Proceedings, 05371 (2005)
52. Wohed, P., Russell, N., ter Hofstede, A.H., Andersson, B., van der Aalst, W.M.: Patterns-based evaluation of open source BPM systems: The cases of jBPM, OpenWFE, and Enhydra Shark. Information and Software Technology 51(8), 1187–1216 (2009)

Artificial Immune System for Forecasting Time Series with Multiple Seasonal Cycles

Grzegorz Dudek

Department of Electrical Engineering, Czestochowa University of Technology,
Al. Armii Krajowej 17, 42-200 Czestochowa, Poland
dudek@el.pcz.czest.pl

Abstract. Many time series exhibit seasonal variations related to the daily, weekly or annual activity. In this paper a new immune inspired univariate method for forecasting time series with multiple seasonal periods is proposed. This method is based on the patterns of time series seasonal sequences: input ones representing sequences preceding the forecast and forecast ones representing the forecasted sequences. The immune system includes two populations of immune memory cells – antibodies, which recognize both types of patterns represented by antigens. The empirical probabilities that the forecast pattern is detected by the kth antibody from the second population while the corresponding input pattern is detected by the jth antibody from the first population, are computed and applied to the forecast construction. The empirical study of the model including sensitivity analysis to changes in parameter values and the robustness to noisy and missing data is performed. The suitability of the proposed approach is illustrated through applications to electrical load forecasting and compared with ARIMA and exponential smoothing approaches.

Keywords: artificial immune system, seasonal time series forecasting, similarity-based methods.

1   Introduction

In general, a time series can be thought of as consisting of four different components: trend, seasonal variations, cyclical variations, and irregular component. The specific functional relationship between these components can assume different forms. Usually they combine in an additive or a multiplicative fashion. Seasonality is defined to be the tendency of time series data to exhibit behavior that repeats itself every n periods. The difference between a cyclical and a seasonal component is that the latter occurs at regular (seasonal) intervals, while cyclical factors have usually a longer duration that varies from cycle to cycle. The presence of the cyclical and multiple seasonal cycles hampers the construction of forecasting models. In this article we concentrate on the seasonal cycles. Seasonal patterns of time series can be examined via correlograms or periodograms based on a Fourier decomposition.

Many economical, business and industrial time series exhibit seasonal behavior. Examples of data with recurrent patterns are: retail sales, industrial production, traffic,

N.T. Nguyen (Ed.): Transactions on CCI XI, LNCS 8065, pp. 176–197, 2013.

weather phenomena, electricity load, calls to call center, gas and water consumption. The recurrent patterns in these cases can be observed within daily, weekly and/or annual periods. Seasonality is often connected with the rhythm of life of the population and its relationship to the variability of the seasons, professional activity, traditions and habits.

A variety of methods have been proposed for seasonal time series forecasting. These include [1]: seasonal ARIMA, exponential smoothing, artificial neural networks, dynamic harmonic regression, vector autoregression, random effect models, and many others. The first three approaches are the most commonly employed to modeling seasonal patterns.

According to Box et al. [2] we can extend the base ARIMA model with just one seasonal pattern for the case of multiple seasonalities. Such an extension we can find in [3]. The inconvenience in the time series modeling using multiple seasonal ARIMA is a combinatorial problem of selecting appropriate model orders.

The basic Holt-Winters exponential smoothing was adapted by Taylor so that it can accommodate two seasonalities [3]. Empirical comparison showed that the resulting forecasts for the new double seasonal Holt-Winters method outperformed those from standard Holt-Winters and also those from a double seasonal ARIMA model. An advantage of the exponential smoothing models is, besides their relative simplicity, that they can be nonlinear. On the other hand it can be viewed as being of high dimension, as it involves initialization and updating of a large number of terms (level, periods of the intraday and intraweek cycles). More parsimonious formulation is proposed in [1]. Recently five new exponentially weighted methods for forecasting time series that consist of both intraweek and intraday seasonal cycles were proposed in [4].

Gould et al. described a state space model developed for the series using the innovation approach which enables to develop explicit models for both additive and multiplicative seasonality [5]. The innovation state space approach provides a theoretical foundation for exponential smoothing methods. This procedure improves on the current approaches by providing a common sense structure to the models, flexibility in modeling seasonal patterns, a potential reduction in the number of parameters to be estimated, and model based prediction intervals.

Artificial neural networks (ANNs), being nonlinear and data-driven in nature, may be well suited to the seasonal time series modeling. One of the major advantage of ANNs, that makes they are so often used in practice, is their great capacity to extract unknown and general information from a given data set even in high-dimensional task. The automated ANN learning releases a designer from the cumbersome procedures of a priori model selection. Although there is another problem: the selection of network architecture as well as the learning algorithm. The most popular ANN type used in forecasting task is multilayer perceptron, which has a property of universal approximation. ANNs are able to directly model seasonality, without the prior seasonal adjustment. An example we can find in [6] where authors conduct a comparative study between ANN and ARIMA models. However Nelson et al. [7] conclude that ANNs trained on deseasonalized data perform significantly better than those with raw data. Zhang and Qi [8] notice that not only deseasonality is important but also detrending. They study the effectiveness of data preprocessing on ANN modeling and forecasting performance. In a large-scale empirical study [9] Zhang and Kline fit a linear trend for detrending and then subtract the estimated trend component from the raw data. For deseasonalizing they employ the method of seasonal index based on centered moving averages.

Decomposition of the time series to cope with seasonalities and trend is a procedure used not only in ANNs, but also in other models, e.g. ARIMA and exponential smoothing. The components showing less complexity than the original time series can be predicted independently and more accurate. A frequently used approach is to decompose the time series on seasonal, trend and stochastic components (e.g. using STL filtering procedure based on LOESS smoother [10]). Other methods of decompositions apply the Fourier transform [11] or the wavelet transform [12]. The simple way to remove seasonality is to define the separate time series for each observation in a cycle, i.e. in the case of cycle of length n, n time series are defined including observations in the same positions in a cycle.

In this paper we propose an approach based on the patterns of the time series seasonal sequences. Using patterns we do not need to decompose a time series. A trend and many seasonal cycles as well as the nonstationarity and heteroscedasticity is not a problem here when using proper pattern definitions. The proposed approach belongs to the class of similarity-based methods [13] and is dedicated to forecasting time series with multiple seasonal periods. The forecast here is constructed using analogies between sequences of the time series with periodicities. An artificial immune system (AIS) is used for detection of similar patterns of sequences. The clusters of patterns are represented by antibodies (AB). Two population of ABs are created which recognize two populations of patterns (antigens) – input ones and forecast ones. The empirical probabilities that the pattern of forecasted sequence is detected by the kth AB from the second population while the corresponding pattern of input sequence is detected by the jth AB from the first population are computed and applied to the forecast construction. This idea is taken from [14], where the Kohonen net was used as a clustering method.

The merits of AIS lie in its pattern recognition and memorization capabilities. The application areas for AIS can be summarized as [15]: learning (clustering, classification, recognition, robotic and control applications), anomaly detection (fault detection, computer and network security applications), and optimization (continuous and combinatorial). Antigen recognition, self-organizing memory, immune response shaping, learning from examples, and generalization capability are valuable properties of immune systems which can be brought to potential forecasting models. The first AIS model dedicated to the time series forecasting was proposed by Dudek [16]. The novelty of the AIS proposed in this paper in the comparison with [16] is that the immune memory is composed of two AB populations and the cross-reactivity thresholds are adapted to learning data during the immune memory creation process.

2   Similarity-Based Forecasting Methods

The similarity-based (SB) methods use analogies between sequences of the time series with seasonal cycles. A course of a time series can be deduced from the behavior of this time series in similar conditions in the past or from the behavior of other time series with similar changes in time. In the first stage of this approach the time series is divided into sequences of length n, which usually contain one seasonal cycle. Fig. 1 shows a periodical time series, where we can observe annual, weekly and daily variations. This series represents hourly electrical loads of the Polish power system. Our task is to forecast the time series elements in the daily period, so the sequences include 24 successive elements of the daily periods.

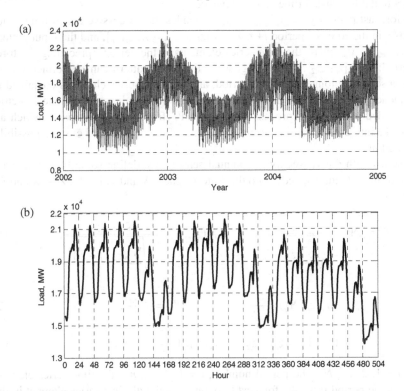

Fig. 1. The load time series of the Polish power system in three-year (a) and three-week (b) intervals

In order to eliminate trend and seasonal variations of periods longer than n (weekly and annual variations in our example), the sequence elements are preprocessed to obtain their patterns. The pattern is a vector with components that are functions of actual time series elements. The input and output (forecast) patterns are defined: $\mathbf{x} = [x_1 \, x_2 \, \ldots \, x_n]^T$ and $\mathbf{y} = [y_1 \, y_2 \, \ldots \, y_n]^T$, respectively. The patterns are paired $(\mathbf{x}_i, \mathbf{y}_i)$, where $\mathbf{y}_i$ is a pattern of the time series sequence succeeding the sequence represented by $\mathbf{x}_i$ and the interval between these sequences (forecast horizon τ) is constant. The SB methods are based on the following assumption: if the process pattern $\mathbf{x}_a$ in a period preceding the forecast moment is similar to the pattern $\mathbf{x}_b$ from the history of this process, then the forecast pattern $\mathbf{y}_a$ is similar to the forecast pattern $\mathbf{y}_b$.

Patterns $\mathbf{x}_a$, $\mathbf{x}_b$ and $\mathbf{y}_b$ are determined from the history of the process. Pairs $\mathbf{x}_a$–$\mathbf{x}_b$ and $\mathbf{y}_a$–$\mathbf{y}_b$ are defined in the same way and are shifted in time by the same number of series elements.

The way of how the $\mathbf{x}$ and $\mathbf{y}$ patterns are defined depends on the time series nature (seasonal variations, trend), the forecast period and the forecast horizon. Functions transforming series elements into patterns should be defined so that patterns carry most information about the process. Moreover, functions transforming forecast sequences into patterns $\mathbf{y}$ should ensure the opposite transformation: from the forecasted pattern $\mathbf{y}$ to the forecasted time series sequence.

The forecast pattern $\mathbf{y}_i = [y_{i,1}\ y_{i,2}\ \dots\ y_{i,n}]$ encodes the successive actual time series elements z in the forecast period $i+\tau$: $\mathbf{z}_{i+\tau} = [z_{i+\tau,1}\ z_{i+\tau,2}\ \dots\ z_{i+\tau,n}]$, and the input pattern $\mathbf{x}_i = [x_{i,1}\ x_{i,2}\ \dots\ x_{i,n}]$ maps the time series elements in the period i preceding the forecast period: $\mathbf{z}_i = [z_{i,1}\ z_{i,2}\ \dots\ z_{i,n}]$. In general, the input pattern can be defined on the basis of a sequence longer than one period, and the time series elements contained in this sequence can be selected in order to ensure the best quality of the model. Vectors $\mathbf{y}$ are encoded using actual process parameters Ψ_i (from the nearest past), which allows to take into consideration current variability of the process and ensures possibility of decoding.

For series with daily, weekly and annual seasons we define some functions mapping the original feature space Z into the pattern spaces X and Y, i.e. $f_x : Z \rightarrow X$ and $f_y : Z \rightarrow Y$:

$$f_x(z_{i,t}, \Psi_i) = \frac{z_{i,t} - \bar{z}_i}{\sqrt{\sum_{l=1}^{n}(z_{i,l} - \bar{z}_i)^2}}, \qquad f_y(z_{i,t}, \Psi_i) = \frac{z_{i+\tau,t} - \bar{z}_i}{\sqrt{\sum_{l=1}^{n}(z_{i+\tau,l} - \bar{z}_i)^2}}, \qquad (1)$$

$$f_x(z_{i,t}, \Psi_i) = \frac{z_{i,t}}{z'}, \qquad f_y(z_{i,t}, \Psi_i) = \frac{z_{i+\tau,t}}{z''}, \qquad (2)$$

$$f_x(z_{i,t}, \Psi_i) = z_{i,t} - z', \qquad f_y(z_{i,t}, \Psi_i) = z_{i+\tau,t} - z'', \qquad (3)$$

where: $i = 1, 2, \dots, N$ – the period number, $t = 1, 2, \dots, n$ – the time series element number in the period i, τ – the forecast horizon, $z_{i,t}$ – the tth time series element in the period i, $\bar{z}_i$ – the mean value of elements in period i, $z' \in \{\bar{z}_i, z_{i-1,t}, z_{i-7,t}, z_{i,t-1}\}$, $z'' \in \{\bar{z}_i, z_{i,t}, z_{i+\tau-7,t}\}$, Ψ_i – the set of coding parameters such as $\bar{z}_i$, z' and z''.

The function f_x defined using (1) expresses normalization of the vectors $\mathbf{z}_i$. After normalization these vectors have the unity length, zero mean and the same variance. When we use the standard deviation of the vector $\mathbf{z}_i$ components in the denominator of equation (1), we receive vector $\mathbf{x}_i$ with the unity variance and zero mean. Note that the nonstationary and heteroscedastic time series is represented by patterns having the same mean and variance.

The components of the x-patterns defined using equations (2) and (3) express, respectively, indices and differences of $z_{i,t}$ and z' or z''.

Forecast patterns are defined using analogous functions to input pattern functions f_x, but they are encoded using the time series elements or characteristics determined from the process history, what enables decoding of the forecasted vector $\mathbf{z}_{i+\tau}$ after the forecast of pattern $\mathbf{y}$ is determined. To calculate the time series element values on the basis of their patterns we use the inverse functions: $f_x^{-1}(x_{i,t}, \Psi_i)$ or $f_y^{-1}(y_{i,t}, \Psi_i)$. For example the inverse functions for (1) are:

$$f_x^{-1}(x_{i,t}, \Psi_i) = x_{i,t} \sqrt{\sum_{l=1}^{n} (z_{i,l} - \overline{z}_i)^2} + \overline{z}_i, \quad f_y^{-1}(y_{i,t}, \Psi_i) = y_{i,t} \sqrt{\sum_{l=1}^{n} (z_{i+\tau,l} - \overline{z}_i)^2} + \overline{z}_i. \quad (4)$$

As a similarity measure between two patterns (real-valued vectors) we can use [17]: the inner product (when the vectors are normalized) or closely related to it cosine similarity measure, Pearson's correlation coefficient or Tanimoto measure. It is useful to define the similarity measures on the base of the distance measures, e.g using linear mapping: $s(\mathbf{x}_a, \mathbf{x}_b) = c - d(\mathbf{x}_a, \mathbf{x}_b)$ or nonlinear mapping: $s(\mathbf{x}_a, \mathbf{x}_b) = 1/(1 + d(\mathbf{x}_a, \mathbf{x}_b))$, where: $s(.,.)$ is a similarity, $d(.,.)$ is a distance and c is a constant greater than the highest value of the distance. The popular distance measures are: Euclidean, Manhatan or Canberra distances. If the components of the vectors are expressed in different units or change in different ranges, in order to offset their impact on the distance, their weighing is recommended.

If for a given time series the statistical analysis confirms the hypothesis that the dependence between similarities of input patterns and similarities between forecast patterns paired with them are not caused by random character of the sample, it justifies the sense of building and using models based on the similarities of patterns of this time series. The statistical analysis of pattern similarities is described in [13].

The forecasting procedure in the case of SB methods can be summarized as follows:

1. Elimination of the trend and seasonal variations of periods longer than n using pattern functions f_x and f_y.
2. Forecasting the pattern $\mathbf{y}$ using similarities between patterns.
3. Reconstruction the time series elements from the forecasted pattern $\mathbf{y}$ using the inverse function f_y^{-1}.

3   Immune Inspired Forecasting Model

The proposed AIS contains immune memory consisting of two populations of ABs. The population of x-antibodies (ABx) detects antigens representing patterns $\mathbf{x} = [x_1, x_2, ..., x_n]^T$ – AGx, while the population of y-antibodies (ABy) detects antigens representing patterns $\mathbf{y} = [y_1, y_2, ..., y_n]^T$ – AGy. The vectors $\mathbf{x}$ and $\mathbf{y}$ form the epitopes of AGs and paratopes of ABs. ABx has the cross-reactivity threshold r defining the AB recognition region. This recognition region is represented by the n-dimensional hypersphere of radius r with center at the point $\mathbf{x}$. Similarly ABy has the recognition region of radius s with center at the point $\mathbf{y}$. The cross-reactivity thresholds are adjusted individually during training. The recognition regions contain AGs with similar epitopes.

AG can be bound to many different ABs of the same type (x or y). The strength of binding (affinity) is dependent on the distance between an epitope and a paratope. AB represents a cluster of similar AGs in the pattern space X or Y. The clusters are overlapped and their sizes depend on the similarity between AGs belonging to them, measured in the both pattern spaces X and Y. The kth ABx can be written as a pair $\{\mathbf{p}_k, r_k\}$, where $\mathbf{p}_k = \mathbf{x}_k$, and the kth ABy as $\{\mathbf{q}_k, s_k\}$, where $\mathbf{q}_k = \mathbf{y}_k$.

After the two population of immune memory have been created, the empirical conditional probabilities $P(ABy_k \mid ABx_j)$, $j, k = 1, 2, \ldots, N$, that the ith AGy stimulates (is recognized by) the kth ABy, when the corresponding ith AGx stimulates the jth ABx, are determined. These probabilities are calculated for each pair of ABx and ABy on the basis of recognition of the training population of AGs.

In the forecasting phase the new AGx, representing pattern $\mathbf{x}^*$, is presented to the trained immune memory. The forecasted pattern $\mathbf{y}$ paired with $\mathbf{x}^*$ is calculated as the mean of ABy paratopes weighted by the conditional probabilities and affinities.

The detailed algorithm of the immune system to forecasting seasonal time series is described below.

```
Training (immune memory creation)
1. Loading of the training populations of antigens.
2. Generation of the antibody populations.
3. Calculation of the cross-reactivity thresholds of
   x-antibodies.
4. Calculation of the cross-reactivity thresholds of
   y-antibodies.
5. Calculation of the empirical conditional probabilities
   P(ABy_k|ABx_j).
Test
6. Forecast determination using y-antibodies, probabili-
   ties P(ABy_k|ABx_j) and affinities.
```

Fig. 2. Pseudocode of the AIS for the seasonal time series forecasting

Step 1. Loading of the training populations of antigens. An AGx represents a single $\mathbf{x}$ pattern, and AGy represents a single $\mathbf{y}$ pattern. Both populations of AGx and AGy are divided into training and test parts in the same way. Immune memory is trained using the training populations, and after learning the model is tested using the test populations.

Step 2. Generation of the antibody populations. The AB populations are created by copying the training populations of AGs (ABs and AGs have the same structure). Thus the paratopes take the form: $\mathbf{p}_k = \mathbf{x}_k$, $\mathbf{q}_k = \mathbf{y}_k$, $k = 1, 2, \ldots, N$. The number of AGs and ABs of both types is the same as the number of learning patterns.

Step 3. Calculation of the cross-reactivity thresholds of x-antibodies. The recognition region of the kth ABx should be as large as possible and cover only the AGx that satisfy two conditions:

(i)   their epitops **x** are similar to the paratope $\mathbf{p}_k$, and

(ii) the AGy paired with them have epitopes **y** similar to the kth ABy paratope – $\mathbf{q}_k$.

The measure of similarity of the ith AGx to the kth ABx is an affinity:

$$a(\mathbf{p}_k,\mathbf{x}_i) = \begin{cases} 0, & \text{if } d(\mathbf{p}_k,\mathbf{x}_i) > r_k \text{ or } r_k = 0 \\ 1 - \dfrac{d(\mathbf{p}_k,\mathbf{x}_i)}{r_k}, & \text{otherwise} \end{cases}, \tag{5}$$

where $d(\mathbf{p}_k, \mathbf{x}_i)$ is the distance between vectors $\mathbf{p}_k$ and $\mathbf{x}_i$, $a(\mathbf{p}_k, \mathbf{x}_i) \in [0, 1]$.
Affinity $a(\mathbf{p}_k, \mathbf{x}_i)$ informs about the degree of membership of the ith AGx to the cluster represented by the kth ABx.

The similarity of the ith AGy to the kth ABy mentioned in (ii) is measured using the forecast error of the time series elements encoded in the paratope of the kth ABy. These elements are forecasted using the epitope of the ith AGy. The forecast error is:

$$\delta_{k,i} = \frac{100}{n} \sum_{t=1}^{n} \frac{| z_{k+\tau,t} - f_y^{-1}(y_{i,t},\Psi_k)|}{z_{k+\tau,t}}, \tag{6}$$

where: $z_{k+\tau,t}$ – the tth time series element of the period $k+\tau$ which is encoded in the paratope of the kth ABy: $q_{k,t} = f_y(z_{k+\tau,t},\Psi_k)$, $f_y^{-1}(y_{i,t},\Psi_k)$ – the inverse function of the pattern **y** returning the forecast of time series element $z_{k+\tau,t}$ using the epitope of the ith AGy.

If the condition $\delta_{k,i} \leq \delta_y$ is satisfied, where δ_y is the error threshold value, it is assumed that the ith AGy is similar to the kth ABy, and ith AGx, paired with this AGy, is classified to class 1. When the above condition is not met the ith AGx is classified to class 2. Thus class 1 indicates the high similarity between ABy and AGy. The classification procedure is performed for each ABx.

The cross-reactivity threshold of kth ABx is defined as follows:

$$r_k = d(\mathbf{p}_k,\mathbf{x}_A) + c[d(\mathbf{p}_k,\mathbf{x}_B) - d(\mathbf{p}_k,\mathbf{x}_A)], \tag{7}$$

where B denotes the nearest AGx of class 2 to the kth ABx, and A denotes the furthest AGx of class 1 satisfying the condition $d(\mathbf{p}_k, \mathbf{x}_A) < d(\mathbf{p}_k, \mathbf{x}_B)$. The parameter $c \in [0, 1)$ allows to adjust the cross-reactivity threshold value from $r_{k\min} = d(\mathbf{p}_k, \mathbf{x}_A)$ to $r_{k\max} = d(\mathbf{p}_k, \mathbf{x}_B)$.

Step 4. Calculation of the cross-reactivity thresholds of y-antibodies. The cross-reactivity threshold of kth ABy is calculated similarly to the above:

$$s_k = d(\mathbf{q}_k,\mathbf{y}_A) + b[d(\mathbf{q}_k,\mathbf{y}_B) - d(\mathbf{q}_k,\mathbf{y}_A)], \tag{8}$$

where B denotes the nearest AGy of class 2 to the kth ABy, and A denotes the furthest AGy of class 1 satisfying the condition $d(\mathbf{q}_k, \mathbf{y}_A) < d(\mathbf{q}_k, \mathbf{y}_B)$. The parameter $b \in [0, 1)$ plays the same role as the parameter c.

The ith AGy is classified to class 1, if for the ith AGx paired with it, there is $\varepsilon_{k,i} \leq \varepsilon_x$, where ε_x is the threshold value and $\varepsilon_{k,i}$ is the forecast error of the time series elements encoded in the paratope of the kth ABx. These elements are forecasted using the epitope of the ith AGx. The forecast error is:

$$\varepsilon_{k,i} = \frac{100}{n} \sum_{t=1}^{n} \frac{| z_{k,t} - f_x^{-1}(x_{i,t}, \Psi_k) |}{z_{k,t}}, \tag{9}$$

where: $z_{k,t}$ – the tth time series element of the period k which is encoded in the paratope of the kth ABx: $p_{k,t} = f_x(z_{k,t}, \Psi_k)$, $f_x^{-1}(x_{i,t}, \Psi_k)$ – the inverse function of pattern x returning the forecast of time series element $z_{k,t}$ using the epitope of the ith AGx.

The ith AGy is recognized by kth ABy if affinity $a(\mathbf{q}_k, \mathbf{y}_i) > 0$, where:

$$a(\mathbf{q}_k, \mathbf{y}_i) = \begin{cases} 0, & \text{if } d(\mathbf{q}_k, \mathbf{y}_i) > s_k \text{ or } s_k = 0 \\ 1 - \dfrac{d(\mathbf{q}_k, \mathbf{y}_i)}{s_k}, & \text{otherwise} \end{cases}, \tag{10}$$

$a(\mathbf{q}_k, \mathbf{y}_i) \in [0, 1]$ expresses the degree of membership of pattern $\mathbf{y}_i$ to the cluster represented by the kth ABy.

Procedure for determining the threshold s_k is thus analogous to the procedure for determining the threshold r_k. The recognition region of kth ABy is as large as possible and covers AGy that satisfy two conditions:
(i)   their epitops $\mathbf{y}$ are similar to the paratope $\mathbf{q}_k$, and
(ii)   the AGx paired with them have epitopes $\mathbf{x}$ similar to the kth ABx paratope – $\mathbf{p}_k$.

This way of forming clusters in pattern space X (Y) makes that their sizes are dependent on the dispersion of y-patterns (x-patterns) paired with patterns belonging to these clusters. Another pattern $\mathbf{x}_i$ ($\mathbf{y}_i$) is appended to the cluster ABx_k (ABy_k) (this is achieved by increasing the cross-reactivity threshold of AB representing this cluster), if the pattern paired with $\mathbf{x}_i$ ($\mathbf{y}_i$) is sufficiently similar to the paratope of the kth ABy_k (ABx_k). The pattern is considered sufficiently similar to the paratope, if it allows to forecast the time series encoded in the paratope with an error no greater than the threshold value. This ensures that the forecast error for the pattern $\mathbf{x}$ ($\mathbf{y}$) has a value not greater than ε_x (δ_y). Lower error thresholds imply smaller clusters, lower bias and greater variance of the model. Mean absolute percentage error (MAPE) here is used as an error measure ((6) and (9)) but other error measures can be applied.

Step 5. Calculation of the empirical conditional probabilities $P(ABy_k|ABx_j)$. After the clustering of both spaces is ready, the successive pairs of antigens (AGx_i, AGy_i), $i = 1, 2, \ldots, N$, are presented to the trained immune memory. The stimulated ABx and

ABy are counted and the empirical frequencies of ABy_k given ABx_j, estimating conditional probabilities $P(ABy_k|ABx_j)$, are determined.

Step 6. Forecast procedure. In the forecast procedure new AGx, representing the pattern $\mathbf{x}^*$, is presented to the immune memory. Let Ω be a set of ABx stimulated by this AGx. The forecasted pattern $\mathbf{y}$ corresponding to $\mathbf{x}^*$ is estimated as follows:

$$\hat{\mathbf{y}} = \sum_{k=1}^{N} w_k \mathbf{q}_k , \tag{11}$$

where

$$w_k = \frac{\displaystyle\sum_{j\in\Omega} P(ABy_k \mid ABx_j)a(\mathbf{p}_j,\mathbf{x}^*)}{\displaystyle\sum_{l=1}^{N}\sum_{j\in\Omega} P(ABy_l \mid ABx_j)a(\mathbf{p}_j,\mathbf{x}^*)} \tag{12}$$

and $\displaystyle\sum_{k=1}^{N} w_k = 1$.

The forecast is calculated as the weighted mean of paratopes $\mathbf{q}_k$. The weights w express the sum of products of the affinities of stimulated memory cells ABx to the AGx and probabilities $P(ABy_k \mid ABx_j)$.

The clusters represented by ABs have spherical shapes, they overlap and their sizes are limited by cross-reactivity thresholds. The number of clusters is here equal to the number of learning patterns, and the means of clusters in the pattern spaces X and Y (paratopes ABx and ABy) are fixed – they lie on the learning patterns.

The cross-reactivity thresholds, determining the cluster sizes, are tuned to the training data in the immune memory learning process. In results the clusters in the space X correspond to compact clusters in the space Y, and vice versa. It leads to more accurate mapping $X \rightarrow Y$. The model has four parameters: the error thresholds (δ_y and ε_x) and the parameters tuning the cross-reactivity thresholds (b and c). Increasing the value of these parameters imply an increase in size of clusters, an increase in the model bias and reduction in its variance.

The training routine is deterministic, which means the fast learning process. The immune memory learning needs only one pass of the training data. The runtime complexity of the training routine is $O(N^2 n)$. The most costly operation is the distance calculation between each ABs and AGs. The runtime complexity of the forecasting procedure is also $O(N^2 n)$.

4   Empirical Study

The described above AIS was applied to the next day electrical load curve forecasting ($\tau = 1$). Short-term load forecasting plays a key role in control and scheduling of power systems and is extremely important for energy suppliers, system operators, financial institutions, and other participants in electric energy generation, transmission, distribution, and markets. Precise load forecasts are necessary for electric companies to make important decisions connected with electric power production and transmission planning, such as unit commitment, generation dispatch, hydro scheduling, hydro-thermal coordination, spinning reserve allocation and interchange evaluation.

The series studied in this section represents the hourly electrical load of the Polish power system from the period 2002-2004. This series is shown in Fig. 1. The time series were divided into training and test parts. The test set contained 30 pairs of patterns from January 2004 (from January 2 to 31) and 31 pairs of patterns from July 2004. The training set contained patterns from the period from January 1, 2002 to the day preceding the day of forecast.

For each day from the test part the separate immune memory was created using the training subset containing AGy representing days of the same type (Monday, ..., Sunday) as the day of forecast and paired with them AGx representing the preceding days (e.g. for forecasting the Sunday load curve, model learns from AGx representing the Saturday patterns and AGy representing the Sunday patterns). This routine of model learning provides fine-tuning its parameters to the changes observed in the current behavior of the time series.

The distance between ABs and AGs was calculated using Euclidean metric. The patterns were defined using (1). MAPE, which is traditionally used in short-term load forecasting, was a forecast error measure.

The model parameters were determined using the grid search method on the training subsets in the local version of the leave-one-out procedure. In this procedure not all patterns are successively removed from the training set to estimate the generalization error but only the k-nearest neighbors of the test x-pattern (k was arbitrarily set to 5). As a result, the model is optimized locally in the neighborhood of the test pattern. It leads to a reduction in learning time.

In the grid search procedure the parameters were changed as follows:

(i) $\delta_y = 1.00, 1.25, ..., 3.00$, $\varepsilon_x = 1.00, 1.25, ..., \delta_y$, at the constant values of $b = c = 1$, and

(ii) $b = c = 0, 0.2, ..., 1.0$, at the optimal values of δ_y and ε_x determined in (i).

It was observed that at lower values of δ_y and c many x-patterns are unrecognized. If $\delta_y \geq 2.25$ and $c = 1$ approximately 99% of the x-patterns are detected by ABx. Increasing δ_y above 2.25 results in increasing the error. Minimum error (MAPE) was observed for $\delta_y = 2.25$, $\varepsilon_x = 1.75$ and $b = c = 1$.

The forecast test MAPE for January was 1.37 and for July was 0.92. Fig. 3 illustrates the empirical conditional probabilities $P(ABy_k|ABx_j)$ estimated on the training set for July 1, 2004. You can observe a specific pattern that indicates which ABs in both populations are activated simultaneously (e.g. the activation of x-antibodies #19–31 corresponds the stronger activation of y-antibodies #19–32, 69–80 and 118–122). These are ABs representing daily cycles lying in the same period of the year. Higher probabilities imply a stronger relationship $X \rightarrow Y$ and greater confidence to the forecast. The weights of activated ABs for this forecasting task are shown in Fig. 4. Here we also can see that the input pattern stimulates ABs representing patterns from the same period of the year as the input pattern. The reconstructed forecast pattern on the background of patterns represented by activated ABs is shown in Fig. 5.

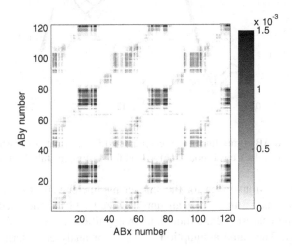

Fig. 3. The empirical conditional probabilities $P(ABy_k|ABx_j)$ estimated on the training set for July 1, 2004

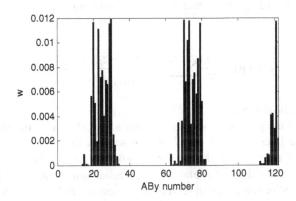

Fig. 4. The weights of activated ABs estimated on the training set for July 1, 2004

4.1   Model Sensitivity to Changes in Parameter Values

The aim of this analysis is to evaluate the influence of the parameter values on the forecasts generated by the model. Fig. 6 shows the test sample forecast errors depending on the parameter values which were changed individually in the ranges: $\varepsilon_x \in [0.5\varepsilon_x^*, 1.5\varepsilon_x^*]$, $\delta_y \in [0.5\delta_y^*, 1.5\delta_y^*]$ and $b = [0.5b^*, \min(1.5b^*, 1)]$, where asterisk denotes the optimal value of the parameter estimated on the training set. It was assumed $c = b$.

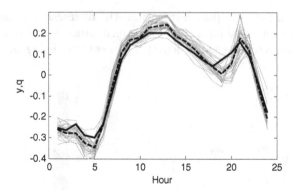

Fig. 5. The activated ABs (gray lines), the reconstructed forecast pattern (dashed line) and the actual forecast pattern (black continuous line) – load forecasting for July 1, 2004

To avoid the situations when for the smaller parameter values many test AGx are unrecognized, it was assumed that an unrecognized AGx is included to the group represented by the nearest ABx, although the recognition region of this ABx does not include the AGx. The same assumption is made for analyses described in Sections 4.2, 4.3 and 5.

Fig. 7 shows the relative percentage difference between the forecasts generated by the optimal model ($\hat{z}_{i+\tau,t}(p^*)$) and the model with non-optimal parameter value ($\hat{z}_{i+\tau,t}(p)$):

$$RPD(p) = \frac{100}{Nn} \sum_{i=1}^{N} \sum_{t=1}^{n} \frac{|\hat{z}_{i+\tau,t}(p) - \hat{z}_{i+\tau,t}(p^*)|}{\hat{z}_{i+\tau,t}(p^*)}. \tag{13}$$

where p denotes the parameter.

The change of the parameter value can cause a step change in the model response which makes the lines in Fig. 6 and 7 are not smooth. From Fig. 6 we can see that overestimation of ε_x and δ_y results in more rapidly increase in the forecast error than underestimation. The values of ε_x and δ_y for the minimum training and test errors are not the same. Smaller values are observed for the test set. RPD achieves 0.3–0.4% for the border values of ε_x and δ_y and only 0.074% for b and c.

The model sensitivity to changes in parameters is defined by the sensitivity index:

$$SI_p = \frac{MAPE_{tst\,max} - MAPE_{tst\,min}}{MAPE_{tst\,min}} \cdot 100. \tag{14}$$

Index (14) informs what is the relative percentage difference between the maximum and minimum errors $MAPE_{tst}$ when the parameter varies in a given range. Its values were: 10.10% for ε_x, 7.42% for δ_y and 2.27% for b and c. Thus the sensitivity to the error thresholds are much greater than for b and c.

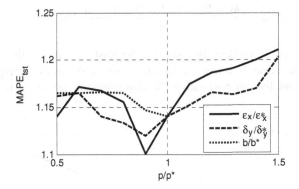

Fig. 6. The forecast error depending on the parameter values

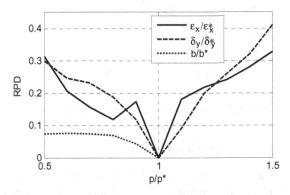

Fig. 7. Relative percentage difference (13) depending on the parameter values

4.2   Model Robustness to Noisy Data

The aim of this analysis is to determine the model robustness to noisy data resulting from the measurement or estimation errors of the time series terms. The noisy patterns are located in the space differently relative to each other than the original patterns. This affects the distances between them, and thus, the probabilities $P(ABy_k|ABx_j)$, affinities and the forecasted pattern.

It is assumed that the actual time series elements are disturbed by random errors:

$$z_l' = z_l \xi_l , \tag{15}$$

where $\xi_l \sim N(1,\sigma)$.

The standard deviation σ was changing in the range from 0 to 0.1. It corresponds to a share of noise in the data $(100 \mid z'-z \mid / z)$ from 0 to 8%. For each σ value 30 training sessions were performed and then mean forecast error was recorded (Fig. 8).

The sensitivity index to noisy data is defined as follows:

$$SI_n(\sigma) = \frac{MAPE_{tst}(\sigma) - MAPE_{tst}(\sigma = 0)}{\sigma} \cdot 100. \tag{16}$$

This index expresses the ratio of the change in forecast error due to the noisy data to the intensity of the noise. The mean value of SI_n for all σ, which corresponds to the slope of the straight line approximating the characteristics presented in Fig. 8, was equal to 65.96%.

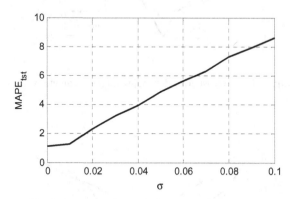

Fig. 8. The forecast error depending on the σ value

4.3   Model Robustness to Missing Data

In this study the robustness to missing input information is analyzed. We assume that some components of the input pattern $\mathbf{x}^*$ are missing. It corresponds to the missing terms of the time series (missing components of vector $\mathbf{z}^*$). In many models (e.g. ARIMA, exponential smoothing, neural networks) this is a serious problem, and the missing data reconstruction is needed. One strategy for dealing with the missing component value is to assign it the value that is most common among training examples. Another strategy is to assign it the values of the corresponding components of the most similar patterns. The proposed AIS copes well with missing components of the input pattern. In such a case the epitopes of AGx and paratopes of ABx are composed of non-missing components. The immune memory creation and test procedures remain unchanged.

When patterns $\mathbf{x}$ and $\mathbf{y}$ are defined using the mean value of elements $\bar{z}_i$, the values of non-missing components are different from their original values. This is caused by different value of $\bar{z}_i$ which is determined now without the missing components. As in the case of noisy data, patterns with missing components locate differently in the space than the original ones, which cause the change of the forecast pattern.

To examine the robustness of the model to missing data we remove m components of the vectors $\mathbf{z}^*$ and then we redefine patterns $\mathbf{x}$ and $\mathbf{y}$. The immune memory is constructed on the training set and the test is performed using $\mathbf{x}^*$ pattern having the same components as $\mathbf{z}^*$. The m components are chosen by random independently for each

$\mathbf{z}^*$. The forecast errors depending on the relative number of missing components are shown in Fig. 9. When the number of missing components is low the deterioration in the model accuracy is not observed. Errors begin to grow rapidly when m exceeds 16.

As a measure of the model sensitivity to the missing components the index SI_m is proposed:

$$SI_m(m) = \frac{MAPE_{tst}(m) - MAPE_{tst}(m=0)}{m/n} \cdot 100. \tag{17}$$

The SI_m value for $m = 6$ was -3.82%, for $m = 12$ was 11.82% and for $m = 18$ was 41.81%.

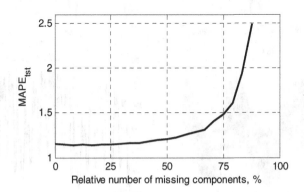

Fig. 9. The forecast error depending on the relative number of missing components

5   Empirical Comparison with Other Models

In this section we compare the proposed AIS with other popular models of the seasonal time series forecasting: ARIMA, exponential smoothing (ES) and double seasonal Holt-Winters (DSHW) method. These models were tested in the next day electrical load curve forecasting problem on three time series of electrical load:

- TS1: time series of the hourly loads of the Polish power system from the period 2002-2004 (this time series was used in analyses described in Section 4). The test sample includes data from 2004 with the exception of 13 untypical days (e.g. holidays).
- TS2: time series of the hourly loads of the local power system from the period July 2001-December 2002. The test sample includes data from the period July-December 2002 except for 8 untypical days.
- TS3: time series of the hourly loads of the local power system from the period 1999-2001. The test sample includes data from 2001 except for 13 untypical days.

Time series TS1 is presented in Fig. 1, TS2 and TS3 in Fig. 10. Time series TS2 and TS3 are more irregular and harder to forecasting than TS1. The measure of the load time series regularity could be the forecast error determined by the naïve method.

The forecast rule in this case is as follows: the forecasted daily cycle is the same as seven days ago. The mean forecast errors, calculated according to this naïve rule, are presented in the last row of Table 1.

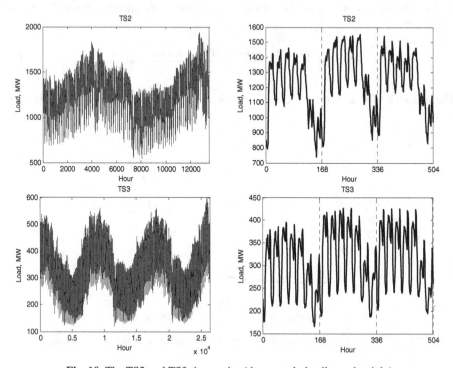

Fig. 10. The TS2 and TS3 time series (three-week details on the right)

In ARIMA the time series were decomposed into 24 series, i.e. for each hour of a day a separate series is created. In this way the daily seasonality is removed. For the independent modeling of these series ARIMA$(p, d, q)\times(P, D, Q)_m$ model is used:

$$\Phi(B^m)\phi(B)(1-B^m)^D(1-B)^d z_t = c + \Theta(B^m)\theta(B)\xi_t, \tag{18}$$

where $\{z_t\}$ is the time series, $\{\xi_t\}$ is a white noise process with mean zero and variance σ^2, B is the backshift operator, $\Phi(.)$, $\phi(.)$, $\Theta(.)$, and $\theta(.)$ are polynomials of order P, p, Q and q respectively, m is the seasonal period (for our data $m = 7$), d and D are orders of nonseasonal and seasonal differencing, respectivelly, and c is a constant.

To find the best ARIMA model for each time series we use a step-wise procedure for traversing the model space which is implemented in the **forecast** package for the **R** system for statistical computing [18]. This automatic procedure returns the model with the lowest Akaike's Information Criterion (AIC) value.

ARIMA model parameters, as well as the parameters of the ES and DSHW models described below, were estimated using 12-week time series fragments immediately preceding the forecasted daily period. Untypical days in these fragments were replaced with the days from the previous weeks.

The ES state space models [19] are classified into 30 types depending on how the seasonal, trend and error components are taken into account. These components can be expressed additively or multiplicatively, and the trend can be damped or not. For example, the ES model with a dumped additive trend, multiplicative seasonality and multiplicative errors is of the form:

$$
\begin{aligned}
\text{Level:} \quad & l_t = (l_{t-1} + \phi b_{t-1})(1 + \alpha \xi_t) \\
\text{Growth:} \quad & b_t = \phi b_{t-1} + \beta(l_{t-1} + \phi b_{t-1})\xi_t \\
\text{Seasonal:} \quad & s_t = s_{t-m}(1 + \gamma \xi_t) \\
\text{Forecast:} \quad & \mu_t = (l_{t-1} + \phi b_{t-1})s_{t-m}
\end{aligned}
\qquad (19)
$$

where l_t represents the level of the series at time t, b_t denotes the growth (or slope) at time t, s_t is the seasonal component of the series at time t, μ_t is the expected value of the forecast at time t, $\alpha, \beta, \gamma \in (0, 1)$ are the smoothing parameters, and $\phi \in (0, 1)$ denotes a damping parameter.

The trend component is a combination of a level term l and a growth term b.

In model (19) there is only one seasonal component. For this reason, as in the case of the ARIMA model, time series is decomposed into 24 series, each of which represents the load at the same hour of a day. These series were modeled independently using an automated procedure implemented in the **forecast** package for the **R** system [18]. In this procedure the initial states of the level, growth and seasonal components are estimated as well as the smoothing and damping parameters. AIC was used for selecting the best model for a given time series.

The DSHW model was proposed by Taylor [3]. This is an exponential smoothing formulation that can accommodate more than one seasonal pattern. This approach does not require the decomposition of the problem. The trend component is treated additively and the seasonal components are treated multiplicatively:

$$
\begin{aligned}
\text{Level:} \quad & l_t = \alpha y_t /(d_{t-m_1} w_{t-m_2}) + (1-\alpha)(l_{t-1} + b_{t-1}) \\
\text{Growth:} \quad & b_t = \beta(l_t - l_{t-1}) + (1-\beta)b_{t-1} \\
\text{Seasonality 1:} \quad & d_t = \delta y_t /(l_t w_{t-m_2}) + (1-\delta)d_{t-m_1} \\
\text{Seasonality 2:} \quad & w_t = \omega y_t /(l_t d_{t-m_1}) + (1-\omega)w_{t-m_2} \\
\text{Forecast:} \quad & \hat{y}_{t+h} = (l_t + hb_t)d_{t-m_1+h}w_{t-m_2+h} + \lambda^h\left(y_t - (l_{t-1} + b_{t-1})d_{t-m_1}w_{t-m_2}\right)
\end{aligned}
\qquad (20)
$$

where d_t and w_t are the seasonal components (daily and weekly in our examples) of the series at time t, $\hat{y}_{t+h}$ is the h step-ahead forecast made from forecast origin t, $\alpha, \beta, \delta, \omega \in (0, 1)$ are the smoothing parameters, m_1 and m_2 are the seasonal periods (for our data $m_1 = 24$ and $m_2 = 168$).

The term involving the parameter λ in (20) is a simple adjustment for first-order autocorrelation. All the parameters: $\alpha, \beta, \delta, \omega$ and λ are estimated in a single procedure by minimizing the sum of squared one step-ahead in-sample errors. The initial smoothed values for the level, trend and seasonal components are estimated by averaging the early observations. These calculations were performed using dshw function from the **forecast** package for the **R** system.

The model selection and training procedures for AIS were the same as described in Section 4. The model was optimized for each test sample, as well as other models.

In Table 1 results of forecasts are presented: MAPE for the test samples and the interquartile range (IQR) of MAPE. The actual and forecasted fragment of TS1 are shown in Fig. 11.

Table 1. Results of forecasting

Model	TS1		TS2		TS3	
	$MAPE_{tst}$	IQR	$MAPE_{tst}$	IQR	$MAPE_{tst}$	IQR
AIS	1.60	1.56	3.35	2.69	3.23	3.10
ARIMA	1.82	1.71	3.41	3.25	3.93	3.68
ES	1.66	1.57	3.16	3.10	3.51	3.23
DSHW	2.23	2.17	3.62	3.45	4.70	4.04
Naïve	3.43	3.42	4.96	3.71	6.62	5.87

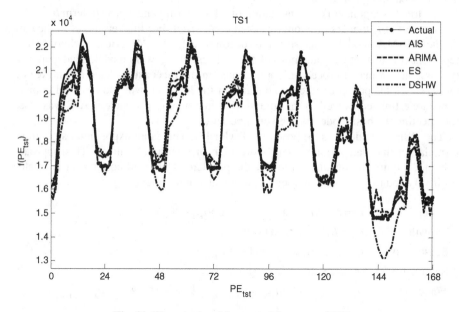

Fig. 11. The actual and forecasted fragment of TS1

In order to indicate the best model we check if the difference between the accuracies of each pair of models is statistically significant using the Wilcoxon rank sum test for equality of medians. The 5% significance level is applied in this study. For only one case: AIS, ES and TS1 the test failed to reject the null hypothesis that errors have the same medians. In all other cases the test indicates the statistically significant difference between errors. Thus the best method for TS1 are AIS and ES, for TS2 is ES, and for TS3 is AIS. The worst method for all cases was DSHW. This model produced several completely incorrect forecasts for TS2 (not included in mean error presented in Table 1). From Table 1 we can see that the naïve method was substantially outperformed by all other methods. It is worth noting as well that the ES model performed better than ARIMA in all cases. It was observed that the ARIMA and ES models produced worst forecasts than AIS and DSHW for the first hours of a day. The proposed AIS was the best model for TS1 and TS3 but not for TS2. This is probably because of the insufficient number of learning points for TS2 (TS2 time series is twice shorter than TS1 and TS3). In this case there are fewer ABs in the immune memory and the local modeling is less accurate.

In Fig. 12 the density functions of the percentage errors (PE) are shown. For ARIMA and ES the shift of the PE densities on the right is observed. It means that these models produced underestimated forecasts.

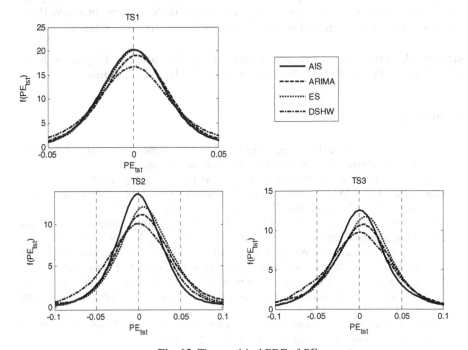

Fig. 12. The empirical PDF of PE_{tst}

6   Conclusions

In this article we deal with forecasting of the seasonal time series which can be non-stationary and heteroscedastic using patterns of the time series seasonal periods. The proposed forecasting method belongs to the class of similarity-based models. These models are based on the assumption that, if patterns of the time series sequences are similar to each other, then the patterns of sequences following them are similar to each other as well. It means that patterns of neighboring sequences are staying in a certain relation, which does not change significantly in time. The more stable this relation is, the more accurate forecasts are. This relation can be shaped by proper pattern definitions and strengthened by elimination of outliers.

The idea of using AIS as a forecasting model is a very promising one. The immune system has some mechanisms useful in the forecasting tasks, such as an ability to recognize and to respond to different patterns, an ability to learn, memorize, encode and decode information.

Unlike other clustering methods used in forecasting models [13], [14], the proposed AIS forms clusters taking into account the forecast error. The cluster sizes are tuned to the data in such a way to minimize the forecast error. Due to the deterministic nature of the model the results are stable and the learning process is rapid. The AIS model also offers robustness to missing data.

The disadvantage of the proposed immune system is limited ability to extrapolation. Regions without the antigens are not represented in the immune memory. However, a lot of models, e.g. neural networks, have problems with extrapolation.

Acknowledgments. The study was supported by the Research Project N N516 415338 financed by the Polish Ministry of Science and Higher Education.

References

1. Taylor, J.W., Snyder, R.D.: Forecasting Intraday Time Series with Multiple Seasonal Cycles Using Parsimonious Seasonal Exponential Smoothing. Department of Econometrics and Business Statistics Working Paper 9/09, Monash University (2009)
2. Box, G.E.P., Jenkins, G.M., Reinsel, G.C.: Time Series Analysis: Forecasting and Control, 3rd edn. Englewod Cliffs, Prentice Hall, New Jersey (1994)
3. Taylor, J.W.: Short-Term Electricity Demand Forecasting Using Double Seasonal Exponential Smoothing. Journal of the Operational Research Society 54, 799–805 (2003)
4. Taylor, J.W.: Exponentially Weighted Methods for Forecasting Intraday Time Series with Multiple Seasonal Cycles. International Journal of Forecasting 26(4), 627–646 (2010)
5. Gould, P.G., Koehler, A.B., Ord, J.K., Snyder, R.D., Hyndman, R.J., Vahid-Araghi, F.: Forecasting Time-Series with Multiple Seasonal Patterns. European Journal of Operational Research 191, 207–222 (2008)
6. Sharda, R., Patil, R.B.: Connectionist Approach to Time Series Prediction: An Empirical Test. Journal of Intelligent Manufacturing 3, 317–323 (1992)
7. Nelson, M., Hill, T., Remus, T., O'Connor, M.: Time Series Forecasting Using NNs: Should the Data Be Deseasonalized First? Journal of Forecasting 18, 359–367 (1999)
8. Zhang, G.P., Qi, M.: Neural Network Forecasting for Seasonal and Trend Time Series. European Journal of Operational Research 160, 501–514 (2005)
9. Zhang, G.P., Kline, D.M.: Quarterly Time-Series Forecasting with Neural Networks. IEEE Transactions on Neural Networks 18(6), 1800–1814 (2007)
10. Cleveland, R.B., Cleveland, W.S., McRae, J.E., Terpenning, I.: STL: A Seasonal-Trend Decomposition Procedure Based on Loess. Journal of Official Statistics 6, 3–73 (1990)
11. Atiya, A., El-Shoura, S., Shaheen, S., El-Sherif, M.: A Comparison Between Neural Networks Forecasting Techniques–Case Study: River Flow Forecasting. IEEE Transactions on Neural Networks 10(2), 402–409 (1999)
12. Soares, L.J., Medeiros, M.C.: Modeling and Forecasting Short-Term Electricity Load: A Comparison of Methods with an Application to Brazilian Data. International Journal of Forecasting 24, 630–644 (2008)

13. Dudek, G.: Similarity-based Approaches to Short-Term Load Forecasting. In: Zhu, J.J., Fung, G.P.C. (eds.) Forecasting Models: Methods and Applications, pp. 161–178. iConcept Press (2010),
http://www.iconceptpress.com/site/download_publishedPaper.php?paper_id=1009170201
14. Lendasse, A., Verleysen, M., de Bodt, E., Cottrell, M., Gregoire, P.: Forecasting Time-Series by Kohonen Classification. In: Proc. the European Symposium on Artificial Neural Networks, pp. 221–226. Bruges, Belgium (1998)
15. Hart, E., Timmis, J.: Application Areas of AIS: The Past, the Present and the Future. Applied Soft Computing 8(1), 191–201 (2008)
16. Dudek, G.: Artificial Immune System for Short-term Electric Load Forecasting. In: Rutkowski, L., Tadeusiewicz, R., Zadeh, L.A., Zurada, J.M. (eds.) ICAISC 2008. LNCS (LNAI), vol. 5097, pp. 1007–1017. Springer, Heidelberg (2008)
17. Theodoridis, S., Koutroumbas, K.: Pattern Recognition, 4th edn. Elsevier Academic Press (2009)
18. Hyndman, R.J., Khandakar, Y.: Automatic Time Series Forecasting: The Forecast Package for R. Journal of Statistical Software 27(3), 1–22 (2008)
19. Hyndman, R.J., Koehler, A.B., Ord, J.K., Snyder, R.D.: Forecasting with Exponential Smoothing: The State Space Approach. Springer (2008)

Machine Ranking of 2-Uncertain Rules
Acquired from Real Data

Beata Jankowska[1] and Magdalena Szymkowiak[2]

[1] Institute of Control and Information Engineering
[2] Institute of Mathematics,
Poznan University of Technology, Pl. M.Sklodowskiej-Curie 5, 60-965 Poznan, Poland
{beata.jankowska,magdalena.szymkowiak}@put.poznan.pl

Abstract. There are many places (e.g. hospital emergency rooms) where reliable diagnostic systems might support people in their work. They could have form of RBSs with uncertainty and use the techniques of forward and backward chaining in their reasoning. The number and the contents of derived hypotheses depend then both on the form of the system's knowledge base and on the inference engine performance. The paper provides detailed considerations on designing and applying particular uncertain rules, namely – 2-uncertain rules. They are equipped with two reliability factors, representing a kind of second order probability. The rules can be acquired from real data of attributive representation. In the paper we propose a method for calculating the two reliability factors. We also suggest how to take advantage of the factors during reasoning, in order to obtain reliable hypotheses. The factors help to rank the rules and to fire them in the best order.

Keywords: attributive data, RBS (Rule-Based System), uncertainty, reliability.

1 Introduction

Nowadays, most of important domain data are stored in repositories connected to some networks. The repositories can be seen as one huge, distributed open database. This is a good starting point for all data integration processes [1] aiming to reduce data incompleteness. It is also a good base for data mining processes [2] that aim to generalize data and extract knowledge from the database.

In fact, the two processes are similar to each other: they consist in joining partial data in aggregate data, with high information contents. Joining them is usually restricted by a number of important constraints, and its result is calculated with the help of special rules. If classical data integration tries to overcome data incompleteness, then data mining aims to increase the level of data certainty. Yet, similar difficulties can be met and should be overcome when performing both processes. Mainly, they are: the diversity of data sources, in particular – in respect of their reliability, and also the diversity of data themselves, both in respect of their syntax (different data representation formats), as well as their semantics (different conceptualisations of the domain). In order to cope with different data reliabilities,

N.T. Nguyen (Ed.): Transactions on CCI XI, LNCS 8065, pp. 198–222, 2013.

one should perform a sophisticated statistical analysis [3]. In turn, syntactic and semantic differences between the data can be overcome by means of schema mapping [4, 5] and path rewriting [6, 7]. In the both data joining processes, the data obtained as a result is uncertain. However, in case of classical data integration – it holds that the greater number of data sources, the higher data uncertainty, whereas in case of data mining – inversely.

1.1 Goal of Research

In the paper, we consider the processes of aggregating data that aim, first to generalize the data and, next to write them in a convenient form. As a rule, the generalized data can be interpreted in a form of implications defining conditional dependencies of conclusions on conjunctions of premises, enriched with factors representing their reliabilities. Such implications can be given a formal shape of association rules or production rules with uncertainty. The obtained rules can form a knowledge base which will be next used for reasoning, with the object of either answering specialized questions or simple reproducing the knowledge of the underlying domain.

Thus, an easy access to a great number of domain data – even distributed or heterogeneous ones – can give rise to an automatic acquisition of reliable, domain knowledge, written in a form of rules. On the one hand, the syntax and semantics of the rules should be adjusted to the form of generalized data obtained by means of aggregation. On the other hand, what is more important, they should guarantee both the effectiveness and the efficiency of reasoning processes performed in the system under construction.

In the paper, we will focus our attention on acquiring such rules for Rule-Based Systems (with Uncertainty) RBSs(U). The main methods of reasoning in RBSs are forward chaining and backward chaining [8]. A forward chaining is being done with a view to reproducing knowledge, whereas a backward chaining – with a view to proving a given hypothesis. A typical RBS use consists in applying forward chaining, optionally preceded by the execution of Magic Transformation [9]. Let us remind that the final results of forward chaining depend – in general – on the order of active rules firing. This order depends, in turn, on the algorithm of agenda conflict resolution which often prefers rules of highest priorities. For this reason, the process of establishing rules' priorities should be carried out very carefully. The more important and reliable the rule is, the higher its priority should be.

Let us observe that most diagnostic procedures and systems have tendency to prefer these relations which express a strong positive dependence between premises and conclusions. Such a requirement is fundamental, among others, for designing association rules [10]. They are created under the constraint that confidence (i.e. the level of probability of rule's conclusions given the occurrence of its premises) is not less than a required threshold. As a rule, weak dependences between premises and conclusions are neglected. The following medical examples prove such an approach not to be justified.

It is common medical practice that before administering a pharmacological treatment, a doctor asks about both the desired and adverse effects of this treatment in patients. The strong effects might make the treatment difficult or even impossible. If taking an oral antibiotic X, used with success in the treatment of sore diseases, causes an adverse effect of torsions in 5% cases only, then one could consider it as an appropriate drug for a physically weakened man with purulent tonsillitis. If this percentage was much higher than reported, then the doctor should think of administering some other drug to the patient, causing less adverse effects compared with the considered antibiotic X. 'The rare occurrence of torsions considered as an effect of taking X' is a kind of weak negative dependence. Observe that, in this situation, 'a frequent no-occurrence of torsions considered as an effect of taking X' is a kind of strong positive dependence.

Let us now consider a medical case similar to the one quoted by Zadeh in his classical paper [11] on fuzzy sets. Assume that in a sickroom a boy, aged 14, suffering from strong vertigo, nausea and spatial orientation disturbances is being seen by a surgeon on duty. If the doctor knows nothing more about him, then the boy will be diagnosed – with probability of about 50% – with a state after 'brain concussion', and – with a similar probability – with 'meningitis'. However, reliable medical evidence confirms the fact that in such a case, in 1% of teenage boys, the correct diagnosis is 'brain tumor'. The diagnosis 'brain tumor' considered as an effect of the observed symptoms is a kind of weak positive dependence. If this dependence was used when diagnosing the boy, it would give the effect of very low probability of 'brain tumor' in him. It is equivalent to high probability of 'no brain tumor', which might be a premise to derive the next important conclusion (e.g. about the necessity to perform further exams with the patient lying supine). The above examples confirm that the rules' priorities should depend on more than strong relationships between the rules' premises and conclusions.

1.2   Related Works

In order to be used in RBSs(U), rules have to implement the notion of 'uncertainty'. The most often used means for expressing uncertainty are ignorance ranges, certainty factors, fuzzy sets and rough sets.

In 1976, Dempster and Shafer (D-S) proposed to model uncertainty by means of not a single factor, but so called 'ignorance range' IR [12]. Their theory was a generalization of the Bayes theory of evidential probability [13]. Put into a rule, IR stands: its left end, called as belief value (BEL) – for the level of unquestionable certainty about the rule's conclusion given the occurrence of the rule's premises; its right end, called as plausibility value (PLS) – for the maximum accessible level of this certainty. Obviously, it holds: $0 \le BEL \le PLS \le 1$. As opposed to the Bayes theory which requires assigning probabilities to all questions of interest, the D-S theory allows to assign BEL values to some from among all possible disjunctions of elementary hypotheses and, next, to carry them over on BEL and PLS values of the question of interest. Besides, an extra rule, called the Dempster's rule of combination [14], says how to combine degrees of belief when they are based on independent items of evidence. It is worth emphasizing that the D-S method performs correctly only when the hypotheses are mutually exclusive. In addition, the Dempster's rule for multiple evidence has apparent exponential-time requirements. In 1981, Barnett [15] showed that the requirements can be reduced to simple polynomial time if the D-S method is applied to the elementary hypotheses and their negations only.

Another important method of modeling uncertainty comes from Shortliffe and Buchanan (S-B). In [16], they proposed to express the level of certainty by means of so called 'certainty factor' CF. This factor informs about the proportional influence of rule's premises on the truthfulness of its conclusion (not about the exact level of conditional probability, but rather about the level of increase of the probability of the rule's conclusion given occurrences of the rule's premises). The value of CF comes from the range <-1; 1>; negative and positive values stand for a negative and positive influence of the rule's premises on the probability of its conclusion, whereas 0 – for the lack of such an influence. In particular, 1 and -1 stand for the conclusion being certainly true and certainly false, respectively. Having assumed a set of input facts, one can reason to conclude different hypotheses and the values of their CFs. A set of inference rules is dedicated to controlling the reasoning, and a set of combination functions – for calculating the mentioned CFs. The B-S model of uncertainty has been implemented in the expert system MYCIN [17] intended to diagnose blood diseases. Although the system has never been used in practice, yet it proposed a correct therapy in about 70 percent of all cases. A syntactically correct reformulation of certainty factor model was given in [18]. Also, a set of combination functions for propagating uncertainty through reasoning chains was redefined there.

The most general method of representing uncertainty is connected with Zadeh's fuzzy logic [19]. In fuzzy rules, premises and conclusions have a form of fuzzy formulas (X is A), where X is a linguistic variable, and A – a linguistic value determined by a fuzzy set on the universe of discourse. When reasoning, fuzzy rules are composed with fuzzy facts (i.e. fuzzy formulas), giving new fuzzy facts as a result. If the expressive power of this method of knowledge representation and reasoning is strong, yet automatic acquiring of fuzzy rules from data in repositories is difficult. In his last paper [20], Zadeh proposes to generalize the notion of fuzziness. Shortly speaking, instead of an earlier fuzzy formula (X is A), he suggests using a formula ((X is A) is B), where B is a linguistic fuzzy value estimating the reliability of (X is A). The proposal is coincident with the database concept of 'data provenance' and its influence on the course and result of data integration processes. The new fuzzy formulas are called Z numbers.

Nowadays, there is a great research interest in rough sets – a formal method to represent uncertainty, coming from Pawlak [21]. The rough set theory is strongly connected with attribute-value data representation. If A stands for a set of attributes, and U for a set of attributive objects under consideration, then each subset $P \subseteq A$ defines an equivalence indiscernibility relation that divides the universe U into classes consisting of objects indiscernible on all attributes from P. These classes can be next used for defining P-lower and P-upper approximations of a set $X \subseteq U$. They are called $\underline{P}X$ and $\overline{P}X$, respectively. The ordered pair $(\underline{P}X, \overline{P}X)$ is called a 'rough set' approximating the extensionally given X, and the difference $\underline{P}X \backslash \overline{P}X$ – as a boundary region of the approximation. It consists of all these objects from U that can but do not necessarily have to belong to the set X. The ratio $\alpha_P(X) = \dfrac{|\underline{P}X|}{|\overline{P}X|}$ stands for the measure of accuracy of the rough set representation of X. Based on the rough set approach, for a chosen set of condition attributes $P \subset A$, and a chosen decision attribute $Q \in A\backslash P$, one can extract from U – a number of decision rules, conditioning the possible decision values of Q on the values of attributes from P. The algorithm uses decision matrices [22], which confront the objects taking the same value on their attributes Q – with the remaining ones. Another important algorithm for extracting rules from the set U is LEM2 [23], that can cope with inconsistent data. Given the lower (or upper) approximation of a decision class, it produces a set of certain (or possible) rules describing the class. The algorithm has proved to be very useful in medical applications [24].

The most recent method of representing uncertainty in rules is called an association rule. The method has been proposed by Agrawal, Imielinski and Swami to discover strong regularities in transactional data stored in large databases. In an association rule, premises and conclusions represent transactions (states, objects, situations) that are often related with one another. Commonly known algorithms of acquiring association rules [10, 25, 26, 27] are based on using factors of support and confidence – the most popular measures of significance and interest. The two factors do not occur in the association rules explicitly – they are hidden. Moreover, they do not differentiate between strong and very strong rules, stating very deep relations between attributes of the objects under consideration. In order to distinguish very strong association rules from a set of all the discovered ones, an extra analysis is needed [28, 29].

The most sophisticated methods of representing uncertainty refer to higher order probabilities [30], e.g. [12, 20]. They enable not only to estimate the chance of satisfying a conclusion given the occurrence of premises, but also to determine the level to which the estimation is reliable. The quality of the method of representing uncertainty depends also on the complexity of operands and operators that are allowed to appear in formulas forming rules' premises and conclusions. An interesting formula concept has been proposed by Attributive Logic with Set Values (over Finite Domains) ALSV(FD) [31].

1.3   Contribution

Our expectations concerning uncertain rules have been as follows:

- the rules are applicable to domains that use both numerical and symbolic values,
- there is a possibility to reason with the rules,
- the rules can be automatically acquired from a real evidence,
- the rules can reflect both simple and complex dependencies of the domain.

The above requirements can be fulfilled better by crisp rules (with appropriate probability factors) rather than fuzzy rules. Among the crisp rules formats mentioned in Section 1.2, the format of decision rules obtained by using rough sets and the format of association rules seem to be most attractive. The strength of decision rules consists in large possibilities of acquiring the rules from complex, attributive data, stored in numerous relational databases of open access. On the other hand, their weak point is weak discrimination between rules' reliabilities (there are only 2 kinds of rules – certain and possible). For a change, association rules express a kind of second order probability. They could have been really distinguished with regard to their reliabilities, if only the factors of confidence and support had been open, not hidden.

Both decision rules and association rules operate on attributes with individual values only. Besides, they are not prepared to reason and reproduce knowledge.

Taking the above into account, we proposed a format of rules with reliabilities [32] which preserves advantages of decision rules and association rules, and tries to eliminate their drawbacks. The format is based on the concept of second order probability. The two public reliability factors used in the rules identify: the level of dependency between rule's components, and the level of rule's quality. The expressive power of the rules is strengthened by enabling literals of set values. It is thanks to them that the rules can reflect dependencies hidden in nested relational databases, which are suitable for storing data from the domains of medicine, geology, geography, etc.

The considered rules can be acquired from (individual or aggregate) attributive data. In [32], an algorithm for acquiring the rules has been proposed. It is based on semantic data integration, and it makes use of some ideas implemented in the commonly known algorithm Apriori [25].

In [33], the mentioned above format of the rule has been extended to the form of so-called 2-uncertain rule. The extension has been done by enabling rule's components (premises and conclusion) to occur in a simple or negated form. The extension made it possible to formulate such rules that were formerly excluded due to negative dependence between their premises and conclusion. Next, it has been shown how to reason correctly and efficiently in a system with 2-uncertain rules [33].

The proposed in [32] algorithm of rules acquisition does not contain a method for calculating rule's reliability factors. As a matter of fact, the functions g_irf and g_grf used in the algorithm are virtual ones. Their implementations were excluded for further analysis. Some important considerations on the reliability factors semantics have been placed in [34]. In this paper, we present a final discussion, and – as an effect – a proposal of implementation of the functions g_irf and g_grf.

1.4   Organization

The contents of the paper is as follows. Section 2 describes a model of RBS with 2-uncertain rules, partially presented in [32]. In that Section, also the problem of propagating uncertainty in a system with such rules is discussed. In particular, a method of managing rules concluding the same conclusion is presented. An algorithm for determining rule's reliability factor irf – the level of certainty of the rule's conclusion given certain occurrences of the rule's premises – is presented in Section 3. Next, an algorithm for determining rule's reliability factor grf – the level of certainty of the rule, taken as a whole, given stored evidence – is presented in Section 4. In Section 5, the theoretical considerations are illustrated with some medical examples of designing and ranking 2-uncertain rules. Finally, Section 6 presents conclusions and directions for future research.

2   Rule-Based Systems with 2-Uncertain Rules

From the point of view of the theory of probability, an RBS with uncertainty can be seen as a set of classical rules enriched with factors representing levels of conditional probabilities. The formats of rules can differ from one another as well in the complexity of formulas used for writing premises and conclusions, as in the form of the attached factors. If rules are extracted from data stored in repositories, then literals in the formulas will come from their domains. The set of acceptable relational and logical operators should be adjusted both to the data semantics and to the domain of system application. The factors dedicated to represent levels of conditional probabilities should refer to first order or second-order probabilities.

According to the proposal from [33], we suggest using rules built from:

- elementary formulas, expressed in a proper subset of ALSV (FD),
- connectives of conjunction, used to join these formulas into complex premises,
- two kinds of crisp reliability factors, used to define internal and global conditional dependencies of the rules.

In the following part of the Section we will shortly present the form of the rule proposed in detail in [33]. We will call it as '2-uncertain rule'. Additionally, we will show how to make a good use of the mentioned reliability factors while reasoning in RBS with such rules in its knowledge base.

2.1   2-Uncertain Rule – Syntax and Semantics

Definition 2.1. Let A stand for a set of concepts from the considered domain (e.g. medicine), chosen arbitrarily in such a way to reflect some part of its reality (e.g. the results of a medical study). Under these assumptions, by 2-uncertain rule R we will mean an implication of the following form:

$$R: \text{ it happens with grf } = p_g$$
$$\text{if } F_1 \text{ and ... and } F_n \tag{1}$$
$$\text{then } F_c \text{ with irf } = p_i$$

where F_k, $1 \leq k \leq n$, and F_c stand for formulas of the form $A_k=Vq_k$ or $\neg(A_k=Vq_k)$ with: $A_k \in A$, called as 'attribute', $\neg$ being the operator of negation, $V_k \subseteq D_k$, called as 'set value', being a subset of domain D_k, i.e. the set of all possible instantiations (values) of A_k, and $q_k \in \{\odot, \oplus\}$, called as 'set qualifier', giving an interpretation for set V_k that should be understood as: $\odot$ – conjunction of all values from V_k, $\oplus$ – disjunction of all values from V_k; grf stands for the name of 'global reliability factor'; irf – for the name of 'internal reliability factor'; p_g, $p_i \in <0, 1>$ – for the values of factors grf and irf, respectively.   □

In 2-uncertain rule (1), formula-conclusion F_c is conditionally dependent on formulas-premises $F_1, ..., F_n$. It is necessarily a positive monotonic dependence. Value p_i of factor irf represents the probability of occurrence of conclusion F_c given certain occurrences of all premises $F_1, ..., F_n$. Besides, 2-uncertain rule (1), taken as a whole, is conditionally dependent on the contents of the database from which it was derived. Value p_g of factor grf represents the probability of that the rule is fully reliable given the database. We think it proper to not let the factor grf directly influence the certainty of conclusion F_c, but let it really influence the course of reasoning.

The format (1) is to be used also for representing 2-uncertain facts (hypotheses), which can be seen as a kind of specific rules, namely – the ones with the obligatory 'true' in the role of premise. So, by 2-uncertain fact H we will mean an implication of the form:

$$H: \text{it happens with grf} = p_{gh}$$
$$\text{if true} \tag{2}$$
$$\text{then } F_c \text{ with irf} = p_{ih}$$

where p_{gh} and p_{ih} have the same syntax and semantics as p_g and p_i from Definition 2.1.

Example 2.1. Here is an exemplary 2-uncertain rule $R1$, drawing a hypothesis on 'brain tumor' for teenage boys diagnosed in the way reported in Section 1.

```
R1: it happens with grf = 0.82
      if Current_symptoms =
            {vertigo, nausea, visual_disturbances}⊙ and
         Gender = {male}⊙ and
         Age_range = {11,…,18}⊕
      then  Diagnosis = {brain_tumor}⊙ with irf = 0.01
```

where the value of irf has been obtained from the proportion $\text{irf}(R1) = \dfrac{3}{305} \cong 0.01$ (3 cases of brain_tumor in 305 teenagers conforming to the criteria of the rule) – see formula (7); and the value of grf – as the following minimum: $\text{grf}(R1) = \min\{0.95, 0.82\}$ – see formula (8), where 0.95 is an approximate value of the rule's

weight: $wg(R1) = \min\left\{1 - 2 \cdot 1.96 \cdot \sqrt{\dfrac{0.01 \cdot (1 - 0.01)}{305}}, 0.95\right\}$ – see formula (10), and

$ac(R1) = 0.82$ is an approximate value of the rule's accuracy – see formula (14). Co-occurrence of the low value of $\text{irf}(R1) = 0.01$ and the high value of $\text{grf}(R1) = 0.82$ deserves underlining. The factors prove that the hypothesis generated by the rule – although specific – is very reliable.   □

2.2   Uncertain Reasoning

A typical RBS(U) manages two kinds of reasoning: forward chaining and backward chaining. As it was said before, we are mainly interested in forward chaining processes. The final result of such reasoning depends as well on the state of RBS(U)'s knowledge base (rules) and the initial state of RBS(U)'s database (axiomatic facts), as on the algorithm of agenda conflict resolution and the functions for propagating uncertainty through reasoning chains. For a RBS(U) with 2-uncertain rules, we think it reasonable to base the choice of active rules from agenda – on rules' priorities which, in turn, should be dependent on the rules' factors grf (see Section 4.4). As regards the functions for propagating uncertainty, all of them but the combination function for multiple production rules concluding the same conclusion will be similar to their equivalents from [18].

The detailed constraints imposed on the process of rules firing will be as follows. Let n 2-uncertain facts with conclusions $F_1, ..., F_n$ be present in RBS(U) database and let they have their grf values equal $p_{g1}, ..., p_{gm}$, respectively, and their irf values equal

$p_{l1},..., p_{ln}$, respectively, at a moment. Then, rule R of the form (1) with $\text{irf}(R)= p_i$ and $\text{grf}(R)= p_g$ will be regarded as active if and only if the relations: $\min\{p_{g1},..., p_{gn}, p_g\} \geq \tau_1$ and $\min\{p_{l1},..., p_{ln}\} \geq \tau_2$ hold, where τ_1 and τ_2 stand for required thresholds of reliabilities. If this rule is situated on the top of the agenda, then, after having it fired, the inference engine will generate a hypothesis H of the form (2), where $p_{gh} = \min\{p_{g1},..., p_{gn}, p_g\}$ and $p_{lh} = \min\{p_{l1},..., p_{ln}\} \cdot p_i$. Should the hypothesis be definitely put into the system database? The answer is: no, it depends on the current state of this database. Namely, the following three cases should be considered:

– there is an axiomatic fact with conclusion F_c in the database,
– there is no fact with conclusion F_c in the database (neither axiomatic, nor derived),
– there is a non-axiomatic fact with conclusion F_c in the database (it means that at least one hypothesis with conclusion F_c has already been derived – see Definition 2.2).

In the first case, the new hypothesis will have no influence on the state of the database – axiomatic facts must not be subjected to any modifications. Next, the second case will result in putting the new, non-axiomatic fact – hypothesis (2) – into the database. In the most complicated third case, a special combination function for multiple rules concluding the same conclusion will be applied.

Before we give a formal definition of this function for RBS(U)s with 2-uncertain rules, let us remind its form used in MYCIN. In that diagnostic system, all multiple rules concluding the same conclusion were taken as equally reliable. Let CF_1 and CF_2 stand for nonnegative certainty factors obtained in two different derivations of some hypothesis H. In MYCIN, the complex certainty factor CF_H of H was then calculated according to the formula: $\text{CF}_H = \text{CF}_1 + (1 - \text{CF}_1) \cdot \text{CF}_2$. It means, that each additional piece of positive evidence contributed in MYCIN to the absolute increase in global certainty factor CF_H. Besides, the final value of CF_H did not depend on the order in which partial CF_j were taken into account (and thus – on the order in which partial hypotheses had been derived). Meanwhile, we postulate that the influence of particular multiple rules concluding the same conclusion on the final value of its factor irf should be differentiated, depending on the importance of the rules.

If we have no knowledge about dependencies in the set of rules, in particular – about the existence of indirect recursion between the rules, then we have to assume that a later rule $(j + i)$, $i>0$, can be – to some extent – influenced by an earlier rule (j) of the derivation, e.g. (j) $F_1{\rightarrow}H$, $(j+1)$ $H{\wedge}F_2{\rightarrow}F_3$, $(j+2)$ $F_3{\wedge}F_4{\rightarrow}H$. Should rules (j) and $(j+2)$ be regarded as equally important? And the two hypotheses H as equally reliable? We think it reasonable to regard the former rule (j) as more important than the latter rule $(j+2)$. As a consequence, we consider rule (j) should have greater effect on the final factor irf of hypothesis H than rule $(j + i)$ has. Since rules are to be fired in the order of decreasing factors grf, then our claim concerning multiple rules concluding the same hypothesis H is as follows: the greater rule's grf, the higher should be the effect of the rule on the final factor irf of H.

To fulfill the above requirement, we will use a special combination function, initially proposed in [34]. Since we expect our inference engine to fire only rules that have high grfs, so we are not so interested in literal grf values as in the order of firing rules that is controlled by these values. Continuing, we propose to put grf values aside and to use fixed weights for calculating the effect of particular rules on the final factor irf. The weights have to be decreasing with the progress of reasoning. The requirement is fulfilled by the function fmr from Definition 2.2.

Definition 2.2. In RBS(U) with 2-uncertain rules, the following function fmr will be used for calculating irf value (p_{ih}) of hypothesis H concluded repeatedly in the same process of reasoning:

$$p_{ih}=fmr(H, \mathbf{KB}, d) \qquad \text{where}$$

$$fmr(H, \mathbf{KB}, j)=\begin{cases} v_1 \cdot p_{i1} & \text{for} \quad j=1 \\ (1-v_j)\cdot fmr(H, \mathbf{KB}, j\text{-}1)+v_j \cdot p_{ij} & \text{for} \quad 2\le j\le d \end{cases} \qquad (3)$$

where **KB** stands for a knowledge base containing 2-uncertain rules; d – for a number of those rules from among all fired in the current process of reasoning, which concluded hypothesis H; j, $j=1,..., d$ – for the ordinal number of a rule concluding hypothesis H; p_{ij}, $j=1,..., d$ – for irf value of that hypothesis H which was derived by the rule of ordinal number j; and v_j, $j=1,..., d$ – for the weight of p_{ij}, calculated from the formula:

$$v_j =\begin{cases} 1 & \text{for} \quad j=1 \\ \dfrac{v_{j\text{-}1}}{t+v_{j\text{-}1}} & \text{for} \quad 2\le j\le d \end{cases} \qquad (4)$$

where t is a constant of proportion between the weights of factors irfs from two successive rules concluding hypothesis H. □

For further considerations, we propose to use $t =1.1$. To realize the influence of factor grf on the process of calculating final factor irf of a hypothesis concluded by multiple 2-uncertain rules, let us analyze the next example of diagnostic reasoning.

Example 2.2. Let us again consider the boy from Example 2.1. Differently than previously, let us assume that the boy complained of a headache. Let us suppose that reasoning is in progress at the moment and none of rules concluding brain_tumor has been fired up to this moment. If rules $R1$ (from Example 2.1) and $R2$ (the one given below) were now successively fired (based on the rules' activities and the advantage of factor grf($R1$)=0.82 over factor grf($R2$)=0.74):

```
R2: it happens with grf = 0.74
      if Current_symptoms =
          {vertigo, visual_disturbances}⊙ and
          Age_range = {10,…,100}⊕
          Anamnesis_result = {headache}⊙
      then Diagnosis = {brain_tumor}⊙ with irf = 0.12
```

then, given certain occurrences of all premises from rules $R1$ and $R2$, after having fired these two rules, the hypothesis on Diagnosis would be as follows ($H1$):

```
H1: it happens with grf = pg12
      if true
      then Diagnosis = {brain_tumor}⊙ with irf = pi12
```

where value p_{g12} is calculated according to the formula proposed in [33]:

$p_{g12}=min\{grf(R1), grf(R2)\}= min\{0.82, 0.74\}=0.74$,

and value p_{i12} – from the formula (3) given above:

$p_{i12}= (1-v_2)\cdot v_1 \cdot 0.01+v_2 \cdot 0.12$. From the formula (4) we obtain:

$$v_2 = \frac{1}{1.1+1} = 0.476,$$

and, finally, we have:

$p_{l12}=(1-0.476)\cdot 1\cdot 0.01+0.476\cdot 0.12 = 0.00524+0.05712 = 0.06236$.

However, if only we swapped these two factors grf and, consequently, reversed the order of firing rules, then the hypothesis on Diagnosis would be as follows (*H2*):

```
H2: it happens with grf = pg21
       if true
       then Diagnosis = {brain_tumor}⊙ with irf = pi21
```

where:

p_{g21}=min{grf(*R2*), grf(*R1*)}=0.74

p_{l21}=(1-0.476)·1·0.12+0.476·0.01 = 0.06288+0.00476 = 0.06764 .

The latter factor irf(*H2*) differs from the former one irf(*H1*) by less than 10% of its value. However, if we used a higher constant of proportion, e.g. t=1.5 or t=2.0, then the difference would be much higher (about 40% and 80%, respectively). Regardless of the value of constant t, factor grf can effectively differentiate 2-uncertain rules with respect to their reliability and importance (see Section 4).   □

3   Internal Rule's Reliability

Our proposal of designing 2-uncertain rules is based on the exploration of data stored in a real evidence base. We assume that these are aggregate data. In medicine, chosen as an exemplary domain, such data characterize not an individual patient but a group of patients. The idea concerning the calculation of reliability factors of rules obtained from individual patients' data was presented in [35]. A detailed description of methods for calculating these factors for rules obtained from aggregate attributive data is the subject of our present considerations. From now on, for abbreviation purposes, we will call such data as tuples.

Definition 3.1. Let **A** stand for a set of attributes, representing concepts from the considered domain (see Definition 2.1). By tuple T_i, we mean the following data format:

$$T_i=<op_1(Ac_1=Vq_{i1})/N_i, ..., op_f(Ac_f=Vq_{if})/N_i, op_{f+1}(ac_1=Vq_{i(f+1)})/N_i, ..., op_{f+g}(ac_g=Vq_{i(f+g)})/N_i,$$
$$op_{f+g+1}(ad_1=Vq_{i(f+g+1)})/L_{i1}, ..., op_{f+g+h}(ad_h=Vq_{i(f+g+h)})/L_{ih}> , \qquad (5)$$

where $Ac_1, ..., Ac_f, ac_1, ..., ac_g, ad_1, ..., ad_h \in$ **A**; in particular, $Ac_1, ..., Ac_f$ stand for attributes that are common-key in tuple T_i; $ac_1, ..., ac_g$ – for attributes that are common-not-key in tuple T_i; $ad_1, ..., ad_h$ – for attributes that are discriminatory in tuple T_i; for each attribute A_k, $A_k \in \{Ac_1, ..., Ac_f, ac_1, ..., ac_g, ad_1, ..., ad_h\}$, formula $op_k(A_k=Vq_{ik})$, with optional (!) operator of negation $op_k=\neg$, represents a pair: attribute A_k and its set value V_{ik}, with q_{ik} standing for the set qualifier; N_i, being a number of individuals 'caught' in aggregate tuple T_i, stands for an attribute's count of each common attribute $A_k \in \{Ac_1, ..., Ac_f, ac_1, ..., ac_g\}$ in tuple T_i, and $L_{i1}, ..., L_{ih}$ stand for attributes' counts of respective discriminatory attributes $ad_1, ..., ad_h$ in tuple T_i [36].   □

Let us assume that tuple T is the final result of the integration (see [36]) of initial tuple T_1 with attached tuples $T_2,\ldots, T_m$, all tuples are of schema (5). Then, from tuple T, the following 2-uncertain rule R can be obtained:

$$R: \text{ it happens with grf } = p_g$$
$$\text{if } F_1: op_1(A_1=Vq_1) \text{ and } \ldots \text{ and } F_n: op_n(A_n=Vq_n) \qquad (6)$$
$$\text{then } F_c:op_c(A_c=Vq_c) \text{ with irf } = p_i$$

where formula F_k: $op_k(A_k=Vq_k)$, $k=1,\ldots, n$, corresponding to common attribute $A_k \in \{Ac_1, \ldots, Ac_f, ac_1, \ldots, ac_g\}$ in T, stands for a premise of rule R, and formula F_c: $op_c(A_c=Vq_c)$, corresponding to chosen discriminate attribute $A_c \in \{ad_1, \ldots, ad_h\}$ in T, stands for a conclusion of rule R. Let us remark that, in extreme situations, set value V_k of common-not-key attribute $A_k \in \{ac_1, \ldots, ac_g\}$ from final tuple T can be equal to:

–   attribute's domain D_k, if :

$q_k=\oplus$ in absence of $op_k=\neg$, or $q_k=\odot$ in presence of $op_k=\neg$;

–   an empty set $\varnothing$, if:

$q_k=\odot$ in absence of $op_k=\neg$, or $q_k=\oplus$ in presence of $op_k=\neg$.

In all these situations formula-premise F_k: $op_k(A_k=Vq_k)$ does not publicly occur in rule R. That is why, we regard common attributes of final tuple T as 'potential' rule's premises.

Definition 3.2. Let us assume that 2-uncertain rule R of schema (6) was obtained from tuple T, being the result of the integration of initial tuple T_1 with tuples $T_2,\ldots, T_m$. Let N_i, $i = 1,\ldots, m$, stand for the attribute's count of common attributes in tuple T_i, and L_i, $i = 1,\ldots, m$ – for the attribute's count of the chosen discriminate attribute of T_i. Then, attribute's count $N(R)$ of common attributes in T (standing for a number of individuals caught in T, and relating to the premises of R) is as follows: $N(R)=\sum_{i=1}^{m}N_i$, and attribute's count $L(R)$ of the chosen discriminate attribute (relating to the conclusion of R) is as follows: $L(R)=\sum_{i=1}^{m}L_i$. By internal reliability factor irf of 2-uncertain rule R we will mean parameter:

$$irf(R)=\frac{L(R)}{N(R)}. \qquad \qquad \square \qquad (7)$$

Factor irf takes value from the range <0; 1> (it is so because in each tuple T_i, $i=1,\ldots, m$, we have: $L_i \leq N_i$). The factor is the counterpart of the confidence from association rules, and also, in statistics – the counterpart of the point estimate of the proportion corresponding to the conditional probability of the rule's conclusion, given the certain occurrence of the rule's premises. High (close to 1) level of irf will be typical of rules expressing a very strong dependence between their premises and conclusion. Obviously, such rules should have high priorities in the knowledge base of RBS. However, we suggest that a low (close to 0) level of irf – typical of a rule expressing a very weak dependence between its premises and conclusion – should also influence a high priority of the rule. Let us remark that a low probability of the fact stated in the rule's conclusion implies a high probability of the opposite one. A rule with extreme irf (close to 1, or close to 0) will be regarded as a rule with a 'characteristic' conclusion. As it is said in Section 4, the rule's priority should be monotonically correlated with factor grf.

4   Global Rule's Reliability

The problem of determining global reliability factor grf seems to be very complex. To solve it correctly, it is necessary to take a few different parameters into account. In our opinion, among others these should be: rule's weight wg and rule's accuracy ac. These parameters will be defined in Sections 4.1 and 4.2, respectively.

Definition 4.1. By global reliability factor grf of 2-uncertain rule R of schema (6) we will mean parameter:

$$grf(R) = \min \{wg(R), ac(R)\}. \qquad\qquad \square \qquad (8)$$

The definition says that a rule is considered to have high global reliability factor grf if it has both a high weight and a high accuracy.

4.1   Rule's Weight

The first parameter that should have an influence on factor grf is rule's weight wg. The value of this parameter will depend mainly on $N(R)$, being the attribute's count of common attributes in final tuple T which is the base for designing rule R.

To determine rule's weight wg, we will firstly estimate $100\% \cdot (1-\alpha)$ confidence interval for factor irf. For a given large enough $N(R)$ ($N(R) \geq 100$, [37]), we can assume that factor irf has an asymptotically standard normal distribution, and the length of its confidence interval can be calculated from the following formula:

$$l_{1-\alpha}(\text{irf}(R)) = 2 \cdot u_{1-\frac{\alpha}{2}} \cdot \sqrt{\frac{\text{irf}(R) \cdot (1 - \text{irf}(R))}{N(R)}} . \qquad (9)$$

We consider it proper to make rule's weight wg dependent on the exactness of irf estimation. This exactness is considered to be the result of subtraction $1 - l_{1-\alpha}(\text{irf})$.

Since factor irf takes value from the range <0; 1>, then the confidence interval of irf should be also limited to the range <0; 1>. In practice, confidence intervals are typically stated at the $1-\alpha=0.95$ confidence level. In such a case, the required critical value of the standard normal distribution is equal $u_{0.975} = 1.96$, and confidence interval length $l_{0.95}(\text{irf})$ takes value from the required range <0; 1> for each $N(R)$ greater or equal 4, $N(R) \geq 4$. (then, obviously, also exactness $1 - l_{0.95}(\text{irf})$ takes values from the range <0; 1>). Taking into account the initial assumption $N(R) \geq 100$, we can define rule's weight wg.

Definition 4.2. By rule's weight wg we will mean parameter:

$$wg(R) = \min\{1 - l_{0.95}(\text{irf}(R)), 0.95\}. \qquad\qquad \square \qquad (10)$$

As we can see, the higher number $N(R)$, the smaller length $l_{0.95}(\text{irf})$, and the higher exactness $1 - l_{0.95}(\text{irf})$. Next, the higher exactness $1 - l_{0.95}(\text{irf})$, the higher weight wg. Hence, we can state that intensive data integration can significantly increase rule's weight wg.

However, let us notice (see formula (9)) that exactness $1- l_{0.95}(\text{irf})$ depends not only on number $N(R)$ but also on the extremity of factor irf: the more extreme (close to 1 or close to 0) factor irf, the smaller length $l_{0.95}(\text{irf})$, and the higher exactness $1-l_{0.95}(\text{irf})$. It means that the more characteristic the rule's conclusion, the higher its weight wg.

The last parameter having an influence on rule's weight wg is the confidence level of the confidence interval estimation. We should remember that the confidence interval includes estimated factor irf with probability not exceeding $1- \alpha = 0.95$. That is why rule's weight wg cannot exceed the value of 0.95.

Let us notice that, in some cases, an increase of the confidence level of the confidence interval estimation can cause an improvement in the rule's weight. The detailed influence of this parameter on the rule's weight will be the subject of our future investigation.

4.2 Rule's Accuracy

The second parameter having an influence on global reliability factor grf is a rule's accuracy. Let us recall that common attributes from a final integrated tuple can be regarded as potential premises of 2-uncertain rule (6) being designed from the tuple, and one chosen discriminatory attribute from the tuple – as the rule's conclusion. As it was emphasized in [34], along with the course of semantic data integration, there can be observed a decrease in the accuracy of a virtual data being designed. A comparison of set values of the same attribute from partial tuples (representing real aggregate data) and from the final integrated tuple (representing virtual data), leads us to determine a relative accuracy of a corresponding formula (premise/conclusion) from 2-uncertain rule under consideration. Furthermore, a comparison of the attribute's set values from the final integrated tuple and from the attribute's domain, enables us to estimate an objective accuracy of the corresponding formula (premise/conclusion) from 2-uncertain rule under consideration. After having estimated the accuracies of all potential rule's premises and its conclusion, we will be able to establish an accuracy of the rule as a whole.

Let us assume that 2-uncertain rule R of schema (6) is being derived from final tuple T, that has been integrated from partial tuples $T_1,\ldots, T_m$, of the forms parallel with (5). First, for each attribute A_k in tuple T we determine its domain D_k of cardinality $|D_k|$ as a set of all the attribute's values from all tuples stored in the considered evidence. Next, we determine set value V_{ki} of attribute A_k in each partial tuple T_i (for $i = 1,\ldots, m$, $V_{ki} \subseteq D_k$). Cardinality $|V_{ki}|$ of set V_{ki} will decide about the accuracy of attribute A_k in tuple T_i. It is clear that attribute A_k has low accuracy in tuple T_i if: cardinality $|V_{ki}|$ is high (close to $|D_k|$) – for $q_k=\oplus$ in absence of $\text{op}_k=\neg$, or $q_k=\odot$ in presence of $\text{op}_k=\neg$, and cardinality $|V_{ki}|$ is low (close to 0, standing for the cardinality of an empty set $\varnothing$) – for $q_k=\odot$ in absence of $\text{op}_k=\neg$, or $q_k=\oplus$ in presence of $\text{op}_k=\neg$. Let us remind, that set value V_k (of cardinality $|V_k|$) of attribute A_k in final integrated tuple T has been obtained, as:

- $V_k = \bigcup_{i=1}^{m} V_{ki}$, in case of: $q_k=\oplus$ in absence of $\text{op}_k=\neg$, or $q_k=\odot$ in presence of $\text{op}_k=\neg$

  (then, we have: $\forall_i |V_k|\geq|V_{ki}|$), or

- $V_k = \bigcap_{i=1}^{m} V_{ki}$, in case of: $q_k=\odot$ in absence of $\text{op}_k=\neg$, or $q_k=\oplus$ in presence of $\text{op}_k=\neg$

  (then, we have: $\forall_i |V_{ki}|\geq |V_k|$).

As it was mentioned above, attributes can decrease their accuracy along with the course of data integration.

Definition 4.3. By parameter $rat(A_{ki})$, enabling to express the relative accuracy of attribute A_k, in partial tuple T_i in comparison to final tuple T, we will mean:

$$rat(A_{ki}) = \begin{cases} \dfrac{|V_{ki}|}{|V_k|} & \text{for } q_k=\oplus \text{ in absence of } op_k=\neg, \text{ or } q_k=\odot \text{ in presence of } op_k=\neg \\[3mm] \dfrac{|V_k|}{|V_{ki}|} & \text{for } q_k=\odot \text{ in absence of } op_k=\neg, \text{ or } q_k=\oplus \text{ in presence of } op_k=\neg \quad (11) \\[3mm] 1 & \text{for } |V_{ki}|=|V_k|=0 . \end{cases}$$
□

This parameter takes values from the range <0; 1>. In particular, it takes maximal value 1 if attribute A_k has the same set values in partial tuple T_i and final tuple T.

Let formula F_k: $op_k(A_k=Vq_k)$ stand for a premise/conclusion from 2-uncertain rule R of schema (6) derived from final integrated tuple T. In order to estimate the relative accuracy of formula F_k we have to pay attention not only to the relative accuracy of attribute A_k in partial tuple T_i, $i = 1,...,m$, but also to maximal attribute's count N_i of tuple T_i (5). The last parameter will decide about the influence of tuple T_i on the relative accuracy of formula F_k.

Definition 4.4. By parameter $raf(F_k)$ determining the relative accuracy of formula F_k: $op_k(A_k=Vq_k)$ we will mean:

$$raf(F_k) = \frac{\sum\limits_{i=1}^{m} N_i \cdot rat(A_{ki})}{N(R)} \qquad (12)$$

where, as it was mentioned in Section 3, $N(R)=\sum\limits_{i=1}^{m} N_i$ stands for the attribute's count of all common attributes in final integrated tuple T.   □

As we can notice, relative accuracy $raf(F_k)$ of formula F_k is defined as a weighted average of relative accuracies $rat(A_{ki})$ of attribute A_k in each partial tuple T_i, $i= 1,..., m$. Relative accuracy $raf(F_k)$ takes value from the range <0; 1>. In particular, it takes maximal value 1 if and only if each relative accuracy $rat(A_{ki})$ of attribute A_k in partial tuple T_i, $i = 1,..., m$, in comparison to final tuple T is equal 1.

Definition 4.5. By parameter $oaf(F_k)$ determining the objective accuracy of formula F_k: $op_k(A_k=Vq_k)$ we will mean:

$$oaf(F_k) = \begin{cases} 1-\dfrac{|V_k|}{|D_k|} & \text{for } q_k=\oplus \text{ in absence of } op_k=\neg, \text{ or } q_k=\odot \text{ in presence of } op_k=\neg \\[3mm] \dfrac{|V_k|}{|D_k|} & \text{for } q_k=\odot \text{ in absence of } op_k=\neg, \text{ or } q_k=\oplus \text{ in presence of } op_k=\neg \end{cases} \qquad (13)$$

where V_k stands for the set value of attribute A_k in final tuple T, and D_k stands for the domain of attribute A_k.   □

The objective accuracy $oaf(F_k)$ of formula F_k takes value from the range $<0; 1>$. It can be high (close to 1) if cardinality $|V_k|$ of set V_k is: low (close to 0) – in case of: $q_k = \oplus$ in absence of $op_k=\neg$, or $q_k=\odot$ in presence of $op_k=\neg$, and high (close to cardinality $|D_k|$) – in case of: $q_k= \odot$ in absence of $op_k=\neg$, or $q_k=\oplus$ in presence of $op_k=\neg$. In such a situation, formula F_k is precisely characterized. For a change, parameter $oaf(F_k)$ might take minimal value 0 if set V_k was equal to: attribute's domain D_k – in case of $q_k = \oplus$ in absence of $op_k=\neg$, or $q_k=\odot$ in presence of $op_k=\neg$, or to an empty set $\varnothing$ – in case of $q_k = \odot$ in absence of $op_k=\neg$, or $q_k=\oplus$ in presence of $op_k=\neg$. Let us recall that in case of $oaf(F_k)=0$, formula F_k will not publicly occur in rule R.

Finally, we can define the accuracy of a rule obtained from final tuple T.

Definition 4.6. Let us assume that z formulas F_k: $op_k(A_k=Vq_k)$, $k=1,...,z$, make a set of all potential premises and one conclusion of 2-uncertain rule R derived from final tuple T. Then, parameter ac determining the accuracy of R can be calculated as the arithmetic mean of: the average of the formulas' relative accuracies and the average of the formulas' objective accuracies:

$$ac(R) = \frac{1}{2}\left(\frac{1}{z}\sum_{k=1}^{z}raf(F_k) + \frac{1}{z}\sum_{k=1}^{z}oaf(F_k)\right). \qquad \square \qquad (14)$$

Parameter ac takes value from the range $<0; 1>$. It can be high (close to 1) if all the considered formulas are of high relative and objective accuracies.

4.3   Factor of Global Rule's Reliability

To conclude, factor $grf(R)=\min\{wg(R), ac(R)\}$ (see formula (8)) depends on the rule's weight and accuracy. When the process of data integration proceeds, the two parameters behave as follows: the rule's weight increases and the rule's accuracy decreases with the progress of data integration. It means that, at the same time, data integration influences factor grf positively (the number of individuals caught in the tuple from which the rule is derived) and negatively (the dispersion of the individuals' characteristics). In order to obtain a high grf value, very sophisticated techniques of data integration should be used.

4.4   Rule's Priority

To summarize, we came to the conclusion that factor $\lfloor grf(R) \cdot 100\rfloor$ (the integral part of $grf(R) \cdot 100$) should be this one deciding the priority of 2-uncertain rule R of schema (6). Obviously, the higher global reliability factor grf, the higher priority of the rule. In case two different rules from the agenda have their factors grf equal, we have to compare factors irf of their premises: the higher the minimal irf of rule's premises $F_1,..., F_n$ ($\min\{p_{i1},..., p_{in}\}$, see Section 2.2), the more reliable the rule. In the light of the above considerations, a high priority rule can be obtained only as a result of integration of huge amounts of data; it should have a characteristic conclusion and premises of high relative and objective accuracies.

5   Some Examples of Ranking Rules with Reliabilities

The following examples will illustrate the designing of 2-uncertain rules, especially the method for the estimation of internal and global rules' reliabilities. In the examples, the values of the estimated parameters are rounded to two decimal places (the rounding is of no importance to the discussion to follow). The data come from a medical database, namely from international clinical trials registers.

5.1   Designing 2-Uncertain Rules

Example 5.1. All the data we consider refer to young patients hospitalized for the bronchial asthma exacerbation [38]. The data report the results of clinical trials carried out on three groups of patients. They can be represented by means of the following tuples of schema (5):

T_1 =< General_Diagnosis={pediatric_asthma}$\odot$/64,
   Drug={short-act_beta2_agonist, inhaled_anticholin}$\odot$/64,
   age_range={0 ,..., 3}$\oplus$/64,
   severity_of_diagn_illness={mild, moderate}$\oplus$/64,
   symptoms={coughing}$\odot$/64
   co_intervention={systemic_corticosteroid}$\odot$/64
   treatment_effects={no_hosp_admission}$\odot$/40,
   ¬(adverse_effects={vomiting}$\odot$)/62>;

T_2 =< General_Diagnosis={pediatric_asthma, pediatric_diabetes}$\odot$/15,
   Drug={short-act_beta2_agonist, inhaled_anticholin}$\odot$/15,
   age_range={1 ,..., 5}$\oplus$/15,
   severity_of_diagn_illness={moderate, severe}$\oplus$/15
   symptoms= {coughing, wheezing}$\odot$/15,
   co_intervention={systemic_corticosteroid}$\odot$/15,
   treatment_effects={no_hosp_admission, stab_of_FEV1, good_sleep_in_night}$\odot$/8,
   ¬(adverse_effects={vomiting}$\odot$)/14>;

T_3 =< General_Diagnosis={pediatric_asthma}$\odot$/140,
   Drug={short-act_beta2_agonist, inhaled_anticholin}$\odot$ /140
   age_range={1 ,..., 4}$\oplus$/140,
   symptoms={coughing}$\odot$/140,
   treatment_effects={no_hosp_admission, stab_of_FEV1}$\odot$/80>
   ¬(adverse_effects={vomiting}$\odot$)/137>.

We assume T_1 to be the initial tuple of an integration process. Then, after having integrated the two tuples T_2 and T_3 with T_1, we obtain the following final integrated tuple T as a result:

T =< General_Diagnosis={pediatric_asthma}$\odot$/219,
   Drug={short-act_beta2_agonist, inhaled_anticholin}$\odot$/219,
   age_range={0 ,..., 5}$\oplus$/219,
   severity_of_diagn_illness={mild, moderate, severe}$\oplus$/219,
   symptoms={coughing}$\odot$/219,
   co_intervention= $\varnothing$ $\odot$/219,
   treatment_effects={no_hosp_admission}$\odot$/128,
   ¬(adverse_effects={vomiting}$\odot$)/213>.

The number of patients caught in final tuple T is equal 219. All of those patients prove the same set values for common-key attributes Ac_1=General_Diagnosis, and Ac_2=Drug, and also for common-not-key attributes ac_1=age_range, ac_2=severity_of_diagn_illness, ac_3=symptoms, and ac_4=co_intervention, whereas 128 from among 219 patients show set value {no_hosp_admission}$\odot$ for discriminatory attribute ad_1=treatment_effects, and 213 from among 219 patients do not show set value {vomiting}$\odot$ for discriminatory attribute ad_2=adverse_effects.

Let us denote successive attributes from tuple T by A_k, k=1,..., 8, and let us assume their domains to be equal:

A_1=Ac_1=General_Diagnosis	D_1={pediatric_asthma, pediatric_diabetes}	$\mid D_1 \mid$=2
A_2=Ac_2=Drug	D_2={short-act_beta2_agonist, inhaled_anticholin}	$\mid D_2 \mid$=2
A_3=ac_1=age_range	D_3={0 ,..., 19}	$\mid D_3 \mid$=20
A_4=ac_2=severity_of_diagn_illness	D_4={mild, moderate, severe}	$\mid D_4 \mid$=3
A_5=ac_3=symptoms	D_5={coughing, wheezing}	$\mid D_5 \mid$=2
A_6=ac_4=co_intervention	D_6={systemic_corticosteroid}	$\mid D_6 \mid$=1
A_7=ad_1=treatment_effects	D_7={no_hosp_admission, stab_of_FEV1, good_sleep_in_night }	$\mid D_7 \mid$=3
A_8=ad_2=adverse_effects	D_8={vomiting}	$\mid D_8 \mid$=1

Let us notice that attributes A_4=severity_of_diagn_illness and A_6=co_intervention do not occur in tuple T_3. We interpret this situation in the following way. Attribute A_4=severity_of_diagn_illness is used in T_3 in absence of negation, and it has its set value qualified by $\oplus$ and equal to the attribute's domain D_4={mild, moderate, severe}, whereas attribute A_6=co_intervention is used in T_3 in absence of negation, and it has its set value qualified by $\odot$ and equal to empty set $\varnothing$. For this reason, also in final tuple T, the set value of A_4=severity_of_diagn_illness is equal to domain D_4, and the set value of A_6=co_intervention is equal to empty set $\varnothing$. As a consequence, formulas corresponding to attributes A_4 and A_6 will not publicly occur in rules designed from tuple T.

As it was mentioned in Section 3, common attributes in tuple T should be regarded as these ones giving potential premises of 2-uncertain rule being designed, while a chosen discriminatory attribute in tuple T – as this one giving a conclusion of the rule. Let us notice that the dependence between potential rule's premises: F_1:General_Diagnosis={pediatric_asthma}$\odot$, F_2:Drug={short-act_beta2_agonist, inhaled_anticholin}$\odot$, F_3:age_range={0,...,5}$\oplus$, F_4:severity_of_diagn_illness={mild, moderate, severe}$\oplus$, F_5:symptoms={coughing}$\odot$, F_6:co_intervention=$\varnothing$ $\odot$, and its possible   conclusions: F_7:treatment_effects={no_hosp_admission}$\odot$ and F_8:¬(adverse_effects={vomiting}$\odot$) is a positive monotonic one (the lower the level of fulfillment of the premises, the lower the level of fulfillment of the conclusion). To conclude, for final integrated tuple T, the following 2-uncertain rules $R3$ and $R4$ of schema (6) can be obtained:

```
R3: it happens with grf = p_g3 :
        if  General_Diagnosis  = {pediatric_asthma}⊙ and
            Drug = {short-act_beta2_agonist, inhaled_anticholin}⊙and
            age_range = {0,…,5}⊕ and
            symptoms = {coughing}⊙
        then treatment_effects = {no_hosp_admission}⊙ with irf = p_i3
```

$R4$: it happens with grf = p_{g4} :
    if General_Diagnosis = {pediatric_asthma}⊙ and
       Drug = {short-act_beta2_agonist, inhaled_anticholin}⊙ and
       age_range = {0,…,5}⊕ and
       symptoms = {coughing}⊙
    then ¬(adverse_effects={vomiting}⊙) with irf = p_{i4} .   □

Further on, we will discuss the calculation of factors irf and grf for the above rules.

5.2   Calculating the Factor of Internal Rule's Reliability

Example 5.2. For exemplary rules $R3$ and $R4$, attributes' counts of those common attributes from T which derived the rules' premises are equal $N(R3)=N(R4)=219$, whereas attributes' counts of those discriminatory attributes from T which derived the rules' conclusions, treatment_effects={no_hosp_admission}⊙ and ¬(adverse_effects= {vomiting}⊙), are equal $L(R3)=128$ and $L(R4)=213$, respectively. Then, internal reliability factors of the rules, calculated from formula (7), are equal:

$$\text{irf}(R3)=p_{i3}=\frac{128}{219}=0.58 \quad \text{and} \quad \text{irf}(R4)=p_{i4}=\frac{213}{219}=0.97 \text{ , respectively.}$$

Let us notice that irf(R3) and irf(R4) have values closer to 1 than to 0. However, while both the rules $R3$ and $R4$ express strong dependencies between premises and conclusion, only one of these dependences can be regarded as a very strong. It is the dependence from rule $R4$ – this one with a characteristic conclusion (see Section 3). In the next Sections, the estimation of factors grf for rules $R3$ and $R4$ will be performed. Let us remind, that factor grf is this one that really determines the rule's priority.   □

5.3   Calculating Rule's Weight

Example 5.3. Now, let us calculate weight wg for exemplary rules $R3$ and $R4$. First, from formula (9), we calculate lengths $l_{0.95}(\text{irf}(R3))$ and $l_{0.95}(\text{irf}(R4))$ of the confidence intervals for factors irf($R3$) and irf($R4$), respectively. The lengths are as follows: $l_{0.95}(\text{irf}(R3))$= 0.13; $l_{0.95}(\text{irf}(R4))$ = 0.04. It means that the exactness of irf from rule $R4$: 1–$l_{0.95}(\text{irf}(R4))$ = 0.96 is higher than the exactness of irf from rule $R3$: 1–$l_{0.95}(\text{irf}(R3))$ = = 0.87. The difference occurs even though rules $R3$ and $R4$ are derived from the same tuple (tuple T, with 219 patients caught inside). The crucial reason for the difference is that the conclusion of rule $R4$ is more characteristic than the conclusion of rule $R3$. Then, by means of formula (10), we determine the rules' weights:

    $wg(R3) = \min\{0.95, 0.87\} = 0.87$ and $wg(R4) = \min\{0.95, 0.96\} = 0.95$.

This means that $R4$ has a higher weight compared to $R3$. It is worth noticing that in case of rule $R3$, the exactness of irf estimation is crucial for its weight, whereas in case of rule $R4$, the confidence level 1–α = 0.95 of the estimation is crucial for its weight.   □

5.4   Calculating Rule's Accuracy

Example 5.4. First, for each attribute A_k of the considered integrated tuple T we determine cardinality $|V_{ki}|$ of set value V_{ki} from partial tuple T_i , $i=1,2,3$, and cardinality $|V_k|$ of set value V_k from this tuple T (see Section 5.1):

for attribute A_1 - $|V_{11}|=1, |V_{12}|=2, |V_{13}|=1,$ $|V_1|=1,$
for attribute A_2 - $|V_{21}|=2, |V_{22}|=2, |V_{23}|=2,$ $|V_2|=2,$
for attribute A_3 - $|V_{31}|=4, |V_{32}|=5, |V_{33}|=4,$ $|V_3|=6,$
for attribute A_4 - $|V_{41}|=2, |V_{42}|=2, |V_{43}|=3,$ $|V_4|=3,$
for attribute A_5 - $|V_{51}|=1, |V_{52}|=2, |V_{53}|=1,$ $|V_5|=1,$
for attribute A_6 - $|V_{61}|=1, |V_{62}|=1, |V_{63}|=0,$ $|V_6|=0,$
for attribute A_7 - $|V_{71}|=1, |V_{72}|=3, |V_{73}|=2,$ $|V_7|=1,$
for attribute A_8 - $|V_{81}|=1, |V_{82}|=1, |V_{83}|=1,$ $|V_8|=1.$

Next, by using formula (11), we can determine for each attribute A_k – the relative accuracy of this attribute in partial tuple T_i, $i=1,2,3$:

for attribute A_1 - $rat (A_{11})=1,$ $rat (A_{12})=0.50,$ $rat (A_{13})=1,$
for attribute A_2 - $rat (A_{21})=1,$ $rat (A_{22})=1,$ $rat (A_{23})=1,$
for attribute A_3 - $rat (A_{31})=0.67,$ $rat (A_{32})=0.83,$ $rat (A_{33})=0.67,$
for attribute A_4 - $rat (A_{41})=0.67,$ $rat (A_{42})=0.67,$ $rat (A_{43})=1,$
for attribute A_5 - $rat (A_{51})=1,$ $rat (A_{52})=0.50,$ $rat (A_{53})=1,$
for attribute A_6 - $rat (A_{61})=0,$ $rat (A_{62})=0,$ $rat (A_{63})=1,$
for attribute A_7 - $rat (A_{71})=1,$ $rat (A_{72})=0.33,$ $rat (A_{73})=0.50,$
for attribute A_8 - $rat (A_{81})=1,$ $rat (A_{82})=1,$ $rat (A_{83})=1.$

Let us remark that only two from among all the attributes, namely A_2 and A_8, did not decrease their accuracies in T compared to any of partial tuples T_1, T_2 and T_3. There is one attribute, namely A_3, that decreased its accuracy in T compared to all of the three partial tuples: in tuple T_1 by 33%, in tuple T_2 by 17%, and in tuple T_3 by 33%. The remaining attributes A_1, A_4, A_5, A_6 and A_7 behaved variously. For instance, attribute A_6 totally decreased its accuracy in T compared to tuples T_1 and T_2, but it retained the whole accuracy compared to tuple T_3.

Now, paying attention to maximal attribute's count N_i of partial tuple T_i, $N_1=64$, $N_2=15$, $N_3=140$, and to maximal attribute's count $N(R)$ of final tuple T, $N(R)=219$, we can estimate, from formula (12), relative accuracy $raf(F_k)$ of each formula F_k (corresponding to attribute A_k, $k=1,...,8$) that could be used in a rule obtained from final tuple T:

$raf(F_1)=0.97, raf(F_2)=1, raf(F_3)=0.68, raf(F_4)=0.88, raf(F_5)=0.97, raf(F_6)=0.64, raf(F_7)=0.63, raf(F_8)=1.$

Let us consider formulas F_7 and F_8. Relative accuracy $raf(F_7)=0.63$ results from the fact that attribute A_7 decreased its accuracy in tuple T compared to two of three partial tuples (T_2 and T_3), whereas relative accuracy $raf(F_8)=1$ results from the fact that attribute A_8 did not decrease its accuracy in T compared to any of partial tuples T_1, T_2, and T_3. Let us remark that, according to the comments concerning the format of 2-uncertain rule (6), formulas F_1, …, F_6 are potential premises for 2-uncertain rule being designed from tuple T, whereas each of two formulas F_7 and F_8 is a conclusion for such a rule, in particular – for rules R3 and R4.

Then, for each formula F_k, $k=1,...,8$, we can determine its objective accuracy $oaf(F_k)$ by means of comparing cardinality $|V_k|$ with cardinality $|D_k|$ (see formula (13)):

$oaf(F_1)=0.50, oaf(F_2)=1, oaf(F_3)=0.70, oaf(F_4)=0, oaf(F_5)=0.50, oaf(F_6)=0, oaf(F_7)=0.33, oaf(F_8)=1.$

As we can see, formulas F_2 and F_8 have 100% of objective accuracy. On the contrary, formulas F_4 and F_6 have 0% of objective accuracy and, consequently, they will not publicly occur in 2-uncertain rules designed from T. Let us, however, remark that the relative accuracies of formulas F_4 and F_6, being equal $raf(F_4)=0.88$ and $raf(F_6)=0.64$, cause the necessity of taking the formulas into account when estimating the accuracy of a corresponding rule designed from tuple T.

Finally, let us estimate the accuracy of our exemplary rules $R3$ and $R4$. From formula (14), we obtain:

$$ac(R3)=\frac{1}{2}\left[\frac{1}{7}(0.97+1+0.68+0.88+0.97+0.64+0.63)+\frac{1}{7}(0.50+1+0.70+0+0.50+0+0.33)\right]=0.63$$

$$ac(R4)=\frac{1}{2}\left[\frac{1}{7}(0.97+1+0.68+0.88+0.97+0.64+1)+\frac{1}{7}(0.50+1+0.70+0+0.50+0+1)\right]=0.70$$

As we can see, the both rules differ only in their conclusions. That is why, the difference in the accuracies of $R3$ and $R4$ is strictly due to the difference in the relative and objective accuracies of the two conclusions. □

5.5   Estimating the Factor of Global Rule's Reliability

Example 5.5. To finish, let us estimate global reliability factors of rules $R3$ and $R4$. From formula (8) we obtain:

$\mathsf{grf}(R3)= p_{g3} = \min \{0.87, 0.63\} = 0.63$ and $\mathsf{grf}(R4)= p_{g4} = \min \{0.95, 0.70\} = 0.70.$

Both in case of rule $R3$ and in case of rule $R4$, the more significant parameter for determining the rule's factor grf is the rule's accuracy. It is the result of the intentional construction of the example: we wished to demonstrate various behaviors of attributes from partial tuples during the process of their integration. For the demonstration, some of the attributes importantly decrease their relative and objective accuracies, giving as a result two potential premises and one conclusion of unusual low relative and objective accuracies (F_4, F_6 and F_7). □

5.6   Completing Exemplary Rules with Reliability Factors

Example 5.6. To sum up, for final integrated tuple T, the following 2-uncertain rules $R3$ and $R4$, with the above determined reliability factors, can be obtained:

```
R3:  it happens with grf = 0.63:
       if  General_Diagnosis  = {pediatric_asthma}⊙ and
           Drug = {short-act_beta2_agonist, inhaled_anticholin}⊙and
           age_range = {0,…,5}⊕ and
           symptoms = {coughing}⊙
       then treatment_effects = {no_hosp_admission}⊙ with irf = 0.58

R4 :  it happens with grf = 0.70:
       if  General_Diagnosis  = {pediatric_asthma}⊙ and
           Drug = {short-act_beta2_agonist, inhaled_anticholin}⊙ and
           age_range = {0,…,5}⊕ and
           symptoms = {coughing}⊙
       then ¬(adverse_effects = {vomiting}⊙) with irf = 0.97 .
```

As we can see, factor grf($R4$) is higher compared to factor grf($R3$). As a result, in the knowledge base of the considered RBS(U), rule $R4$ will have a higher priority than rule $R3$ (see Section 4.4). While it is true that both the rules have identical premises, nevertheless, they have different conclusions. The difference in the levels of the both grfs is a consequence of the two following facts:

– the conclusion of $R4$ is more characteristic than the conclusion of $R3$, and,
– both the relative and objective accuracy of the conclusion of $R4$ is higher compared to its counterpart for the conclusion of $R3$.   □

6   Conclusions

In the paper we have proposed a method for ranking 2-uncertain rules obtained from real evidence data. The method can facilitate creating high quality diagnostic RBSs(U). The rules are acquired from aggregate attributive data with set values, by means of their semantic integration. The format of 2-uncertain rules is based on using two reliability factors – irf and grf. Each of the two represents a kind of conditional probability, namely: irf – probability of the rule's conclusion given the (certain) occurrence of the rule's premises; grf – reliability of the rule as a whole, given the evidence from which it was derived. Additionally, the rule's priority is linearly dependent on factor grf.

Internal reliability factor irf is the ratio of the number of individual data (the data concerning single objects or situations) that satisfy both the rule's premises and the rule's conclusion to the number of those ones that satisfy the rule's premises. At the stage of designing 2-uncertain rules, factor irf is calculated as the first of the two rule's factors. Next, it influences, to some extent, factor grf: the more extreme factor irf (close to 1 or close to 0), the higher factor grf. When forward chaining, factor irf of a fired rule influences factor irf of each hypothesis having this rule on its derivation path.

At the stage of designing 2-uncertain rules, global reliability factor grf is calculated as dependent on the number of individual data that have been integrated in the final aggregate data: the greater the number, the higher factor grf. It also depends on the quantitative differences between those individual data: the higher differences, the smaller factor grf. In order to examine the level of the mentioned differences, one should have additional knowledge of the cardinalities of data attributes' domains. Obviously, factor grf of a fired rule influences factor grf of a hypothesis generated by this rule. Besides, the rule's priority, which is linearly dependent on the rule's factor grf, decides on the order of firing active rules from the agenda. In such a way, it can indirectly influence factors irf of newly derived hypotheses (see the combination function for multiple rules concluding the same conclusions).

As it was said before, 2-uncertain rules can be used with success in diagnostic RBSs(U), in particular – in systems to support medical diagnostics. In general, medical diagnostic actions can be classified in two categories:

- initial actions, which result in drawing a general picture of patient's condition (they include anamnesis – if possible, and some physical examinations, e.g. blood pressure measurement, temperature measurement, bronchial auscultation, heart auscultation),
- advanced actions, which consist in verifying different diagnostic hypotheses.

In the first case, a RBS(U) with 2-uncertain rules can be used to generate as many medical conclusions as possible from the facts confirmed when initial diagnosing (the mode of forward chaining). Factors grf help to rank the rules and to fire them in the best order. Optionally, the system can constrain either the threshold of reliability of derived conclusions, or the maximum number of those conclusions (remark, that successively generated conclusions have their $grfs$ smaller and smaller). The conclusions will be derived together with their probabilities (factors irf).

In the second case, the RBS(U) with 2-uncertain rules can be used to advise with a concrete hypothesis: is it no less/more probable than we would expect it to be? Then, factors irf and grf can be applied to optimize the process of searching for an answer (the mode of backward chaining).

The proposed strategies for qualification 2-uncertain rules to the agenda and for resolving conflicts in the agenda are simple ones. On request, they can be modified. Thanks to the existence of two reliability factors in rules, one can think of different sophisticated strategies suitable for use in RBSs(U) with such rules.

The algorithm for designing 2-uncertain rules from attributive data with set values [32] makes good use of taxonomical relations between concepts from the considered domain. These relations can next be used when reasoning, in order to enable better matching between assumed facts and the rules' premises. Let us remark, that the taxonomical relations (and, possibly, other ontological relations) could be also used for better estimation of the similarity between corresponding set values from integrated tuples. At present, the set values are examined from the point of view of their cardinalities. In the future, we are going to introduce the notion of taxonomical distance between concepts from the domain. Based on that distance, we will define qualitative differences between set values, and next between attributive data. The new metrics could give better estimation of the rule's accuracy.

In the paper, we consider aggregate attributive data. In the examples, they represent medical results of not individual patients but of groups of patients. Obviously, the presented method of ranking rules can be easily adapted to individual attributive data too.

The proposed format of 2-uncertain rule, together with the set of combination functions for propagating uncertainty, makes up a model for such RBSs(U) that are capable of performing both forward and backward chaining. The effectiveness of the model will depend as well on the number and quality of rules in the system's knowledge base, as on detailed constraints imposed on the course of reasoning.

It follows from the combinatorial analysis, that the number of 2-uncertain rules that might be acquired from 1000 aggregate attributive data with 5 non-key attributes on average, can reach 80 000. After having deleted redundant rules and inconsistent pairs or rules, there can be still several thousand of rules in the system's knowledge base. In view of that, there should be proposed different varieties of reasoning modes which will enable to obtain, depending on need, a great number of tolerably reliable hypotheses, or a small number of highly reliable hypotheses. Factor grf can be used to solve this problem without obstacles.

The proposed RBS(U) with 2-uncertain rules needs thorough testing, both in respect of effectiveness and efficiency. Before the testing procedure is performed, first the metrics for 'correctness of the system's answer' should be defined.

Acknowledgments. The research has been supported by PUT under grant DS 45-083/13.

References

1. Lenzerini, M.: Data Integration: A Theoretical Perspective. In: Popa, L. (ed.) Proc. 21st ACM Symposium on Principles of Database Systems, pp. 233–246. ACM, Madison (2002)
2. Hand, D., Mannila, H., Smyth, P.: Principles of Data Mining. The MIT Press (2001)
3. Rossman, A.J., Chance, B., von Oehsen, B.J.: Workshop Statistics: Discovery with Data and the Graphic Calculator, 3rd edn. John Wiley & Sons (2008)
4. Fagin, R., Kolaitis, P.G., Miller, R.J., Popa, L.: Data Exchange: Semantics and Query Answering. Theoretical Computer Science 336(1), 89–124 (2005)
5. Fagin, R., Kolaitis, P.G., Popa, L., Tan, W.C.: Schema Mapping Evolution through Composition and Inversion. In: Bellahsene, Z., Bonifati, A., Rahm, E. (eds.) Schema Matching and Mapping, pp. 191–222. Springer (2011)
6. Gottlob, G., Koch, C., Pichler, R.: Efficient algorithms for processing XPath queries. In: Bernstein, P.A., Ioannidis, Y.E., Ramakrishnan, R., Papadias, D. (eds.) Proc. 28th Int. Conf. on Very Large Data Bases, pp. 95–106. Morgan Kaufmann, Hong Kong (2002)
7. XML Path Language (XPath) 2.0, 2nd edn. (2010),
   `http://www.w3.org/TR/xpath20/`
8. Ligęza, A.: Logical Foundations for Rule-Based Systems, 2nd edn. Springer, Heidelberg (2006)
9. Beeri, C., Ramakrishnan, R.: On the Power of Magic. Journal of Logic Programming 10, 255–299 (1991)
10. Agraval, R., Imielinski, T., Swani, A.: Mining association rules between sets of items in large databases. SIGMOD Record 22(2), 805–810 (1993)
11. Zadeh, L.: Fuzzy sets. Information and Control 8, 338–353 (1965)
12. Shafer, G.: A Mathematical Theory of Evidence. Princeton University Press (1976)
13. Bayes, T.: An Essay towards solving a Problem in the Doctrine of Chances. Philosophical Transactions of the Royal Society of London 53, 370–418 (1763)
14. Dempster, A.P.: A generalization of Bayesian inference. Journal of the Royal Statistical Society, Series B 30, 205–247 (1968)
15. Barnett, J.A.: Computational methods for a mathematical theory of evidence. In: Proc. 7th Int. Joint Conf. on Artificial Intelligence, Vancouver, pp. 868–875 (1981)
16. Shortliffe, E.H.: Computer-Based Medical Consultations: MYCIN. Elsevier/North Holland, New York (1976)
17. Buchanan, B.G., Shortliffe, E.H.: Rule-Based Expert Systems: The MYCIN Experiments of the Stanford Heuristic Programming Project. Addison Wesley, Reading (1984)
18. Van der Gaag, L.C.: A conceptual model for inexact reasoning in rule-based systems. Int. Journal of Approximate Reasoning 3(3), 239–258 (1989)
19. Zadeh, L.A.: Fuzzy sets as a basis for a theory of possibility. Fuzzy Sets and Systems 1(1), 3–28 (1978)
20. Zadeh, L.A.: A Note on Z-numbers. Information Sciences 181(14), 2923–2932 (2011)

21. Pawlak, Z.: Rough sets. Int. Journal of Parallel Programming 11(5), 341–356 (1982)
22. Ziarko, W., Shan, N.: Discovering attribute relationships, dependencies and rules by using rough sets. In: Proc. 28th Annual Hawaii Int. Conf. on System Sciences, Hawaii, pp. 293–299 (1995)
23. Grzymala-Busse, J., Wang, A.: Modified algorithms LEM1 and LEM2 for rule induction from data with missing attribute values. In: Proc. 5th Int. Workshop on Rough Sets and Soft Computing, pp. 69–72 (1997)
24. Ilczuk, G., Wakulicz-Deja, A.: Rough Sets Approach to Medical Diagnosis System. In: Szczepaniak, P.S., Kacprzyk, J., Niewiadomski, A. (eds.) AWIC 2005. LNCS (LNAI), vol. 3528, pp. 204–210. Springer, Heidelberg (2005)
25. Agrawal, R., Srikant, R.: Fast algorithms for mining association rules in large databases. In: Bocca, J.B., Jarke, M., Zaniolo, C. (eds.) Proc. 20th Int. Conference on Very Large Data Bases, pp. 487–499. Morgan Kaufmann, Santiago de Chile (1994)
26. Zaki, M.J.: Scalable algorithms for association mining. IEEE Transactions on Knowledge Data Engineering 12(3), 372–390 (2000)
27. Han, J., Pei, J., Yin, Y., Mao, R.: Mining frequent patterns without candidate generation. Data Mining Knowledge Discovery 8, 53–87 (2004)
28. Berzal, F., Blanco, I., Sánchez, D., Vila, M.-A.: A new framework to assess association rules. In: Hoffmann, F., Adams, N., Fisher, D., Guimarães, G., Hand, D.J. (eds.) IDA 2001. LNCS, vol. 2189, pp. 95–104. Springer, Heidelberg (2001)
29. Baqui, S., Just, J., Baqui, S.C.: Deriving strong association rules using a dependency criterion, the lift measure. International Journal of Data Analysis Techniques and Strategies 1(3), 297–312 (2009)
30. Gaifman, H.: A Theory of Higher Order Probabilities. In: Proc. 1st Conf. on Theoretical Aspects of Reasoning about Knowledge, pp. 275–292. Morgan Kaufmann, Monterey (1986)
31. Nalepa, G.J., Ligeza, A.: On ALSV Rules Formulation and Inference. In: Lane, H.C., Guesgen, H.W. (eds.) Proc. 2nd Int. FLAIRS Conference, pp. 396–401. AAAI Press, Florida (2009)
32. Jankowska, B.: Using Semantic Data Integration to Create Reliable Rule-based Systems with Uncertainty. Engineering Applications of Artificial Intelligence 24(8), 1499–1509 (2011)
33. Jankowska, B.: Evidence-based model for 2-uncertain rules and inexact reasoning. Medical Informatics & Technologies 20, 39–47 (2012)
34. Jankowska, B., Szymkowiak, M.: On Ranking Production Rules for Rule-Based Systems with Uncertainty. In: Jędrzejowicz, P., Nguyen, N.T., Hoang, K. (eds.) ICCCI 2011, Part I. LNCS, vol. 6922, pp. 546–556. Springer, Heidelberg (2011)
35. Szymkowiak, M., Jankowska, B.: Discovering Medical Knowledge from Data in Patients' Files. In: Nguyen, N.T., Kowalczyk, R., Chen, S.-M. (eds.) ICCCI 2009. LNCS, vol. 5796, pp. 128–139. Springer, Heidelberg (2009)
36. Jankowska, B., Szymkowiak, M.: Designing Medical Production Rules from Semantically Integrated Data. Journal of Medical Informatics & Technologies 16, 95–102 (2010)
37. Krysicki, W., et al.: Mathematical Statistics. PWN, Warszawa (2006) (in Polish)
38. Plotnick, L.H., Ducharme, F.M.: Combined Inhaled Anticholinergics and Beta2-Agonists for Initial Treatment of Acute Asthma in Children. The Cochrane Library (2005)

Author Index